Composite Materials Technology

Neural Network Applications

Composite Materials Technology

Neural Network Applications

Edited by S.M. Sapuan and I.M. Mujtaba

CRC Press
Taylor & Francis Group
Boca Raton London New York

CRC Press is an imprint of the
Taylor & Francis Group, an **informa** business

composite material technology from damage detection, design and analysis, mechanical properties, to materials and process selection.

Twelve chapters have been compiled and edited as a result of the contributions of various authors from various parts of the world, such as the United States, United Kingdom, Italy, Brazil, Australia, Malaysia, Bangladesh, Turkey, United Arab Emirates, and Indonesia. The book is divided into four parts. Part 1 gives the introduction and a review of literature in the area of ANN in composite materials technology. In Part 2, five chapters are included, and all the chapters are grouped under the common theme of structural health monitoring. Mechanical properties are reported in Part 3 where it comprises three chapters. Finally, design, analysis, and materials selection are presented in three papers in Part 4.

Acknowledgments

Alhamdulillah—all praises to almighty Allah who made it possible for the editors to complete this book.

The editors gratefully acknowledge the Universiti Putra Malaysia, which provided financial support to S. M. Sapuan during his sabbatical leave as visiting academic at the School of Engineering, Design and Technology, University of Bradford, UK, in 2007. During this visit, this book was initiated.

This book includes contributions from the United States, United Kingdom, Malaysia, Italy, Brazil, Australia, Indonesia, United Arab Emirates, Turkey, and Bangladesh. The editors are indebted to all the contributors who worked hard to produce the manuscripts.

The editors would like to express sincere gratitude to the reviewers who made enormous efforts to review each manuscript and provide useful comments.

The editors gratefully acknowledge the editorial assistance provided by Mr. Mohd Zuhri Mohamed Yusoff and Mr. Mohamad Ridzwan Ishak, master of science students at the Universiti Putra Malaysia.

The editors would like to express their appreciation to CRC Press, Boca Raton, for publishing this book, particularly to Ms. Jennifer Ahringer and Ms. Allison Shatkin.

S. M. Sapuan would like to thank the support and motivation given by his wife Nadiah Zainal Abidin, his daughter Qurratu Aini, and his mother Rogayah Wagimon during the preparation of this book. Similarly, I. M. Mujtaba would like to thank his wife Nasreen and his children Summayya, Maria, Hamza, and Usama for their great support and continuous encouragement.

Editors

S. M. Sapuan is a professor of composite materials and the head of the Department of Mechanical and Manufacturing Engineering, Universiti Putra Malaysia (UPM). He is the vice president and honorary member of Asian Polymer Association; fellow of Institute of Materials, Malaysia; life fellow, International Biographical Association; life member, Institute of Energy, Malaysia; member, Society of Automotive Engineers International; member, International Association of Engineers; member, Plastics and Rubber Institute, Malaysia; and a professional engineer. He has published more than 200 papers in refereed journals, more than 200 papers in conferences/seminars, and six books on engineering. He also holds three Malaysian patents. Professor Sapuan's research interests include automotive composites, concurrent engineering, engineering design methods, natural fiber composites and neural network, and expert system in composite materials selection. He is an editor of a special issue on Composite Materials Technology in the *American Journal of Applied Sciences*. In addition, he has edited 11 monographs. He sits on editorial boards for 18 journals and research bulletins. He has reviewed more than 140 papers for refereed journals. He is the recipient of the Excellence Putra Publication Award, UPM Excellence Award from Science Publication, New York, UPM Excellence Researcher Award in Journal Publication, UPM Vice Chancellor Fellowship Prize, and ISESCO Science Prize, Morocco.

I. M. Mujtaba is a professor of computational process engineering in the School of Engineering, Design and Technology at the University of Bradford, UK. He is a fellow of the IChemE, a chartered chemical engineer, and a chartered scientist. Professor Mujtaba is actively involved in many research areas like dynamic modeling, simulation, optimization, and control of batch and continuous chemical processes with specific interests in distillation, industrial reactors, refinery processes, and desalination. He has published more than 110 technical papers in major engineering journals, international conference proceedings, and books. He is a coeditor of the book *Application of Neural Networks and Other Learning Technologies in Process Engineering* published by the Imperial College Press, London, in 2001 (http://www.icpress.co.uk/books/compsci/p225.html). Also, he is the author of the book *Batch Distillation: Design & Operation* published by the Imperial College Press, London, in 2004 (http://www.icpress.co.uk/books/engineering/p319.html).

Contributors

Y. Al-Assaf
Department of Mechanical
    Engineering
College of Engineering
American University of Sharjah
Sharjah, UAE

A. C. Ancelotti, Jr.
Departamento de Química
Instituto Tecnológico de
    Aeronáutica
São José dos Campos, São Paolo,
    Brazil

M. K. Apalak
Department of Mechanical
    Engineering
Erciyes University
Kayseri, Turkey

E. M. Bezerra
Departamento de Química
Instituto Tecnológico de
    Aeronáutica
São José dos Campos, São Paolo,
    Brazil

C. A. R. Brito, Jr.
Departamento de Química
Instituto Tecnológico de
    Aeronáutica
São José dos Campos, São Paolo,
    Brazil

M. Demetgul
Technical Education Faculty
Marmara University
Goztepe, Istanbul, Turkey

T. D'Orazio
Institute of Intelligent Systems for
    Automation
CNR
Bari, Italy

H. El Kadi
Department of Mechanical
    Engineering
College of Engineering
American University of Sharjah
Sharjah, UAE

C. Guaragnella
Politecnico di Bari
DEE
Bari, Italy

M. Hasan
Department of Materials and
    Metallurgical Engineering
Bangladesh University of
    Engineering and Technology
Dhaka, Bangladesh

I. Herszberg
Cooperative Research Centre for
    Advanced Composite Structures
    (CRC-ACS)
Fishermans Bend, Victoria,
    Australia

M. I. P. Hidayat
Department of Materials and
    Metallurgy Engineering
Faculty of Industrial Technology
Institute of Technology Sepuluh
    Nopember
Surabaya, East Java, Indonesia

M. E. Hoque
Department of Mechanical,
  Materials and Manufacturing
  Engineering
University of Nottingham Malaysia
  Campus
Semenyih, Selangor, Malaysia

S. John
School of Aerospace, Mechanical
  and Manufacturing Engineering
RMIT University
Bundoora East Campus
Bundoora, Victoria, Australia

A. Kesavan
National Australian Pipelines
Whittlesea, Victoria, Australia

M. Leo
Institute of Intelligent Systems for
  Automation
CNR
Bari, Italy

S. Mahzan
Faculty of Mechanical and
  Manufacturing Engineering
Universiti Tun Hussein Onn
  Malaysia
Batu Pahat, Johor, Malaysia

G. Manson
Department of Mechanical
  Engineering
University of Sheffield
Sheffield, UK

I. M. Mujtaba
School of Engineering, Design and
  Technology
University of Bradford
West Yorkshire, UK

F. Mustapha
Department of Aerospace
  Engineering
Universiti Putra Malaysia
Serdang, Selangor, Malaysia

L. C. Pardini
Departamento de Química
Instituto Tecnológico de
  Aeronáutica
São José dos Campos, São Paolo,
  Brazil

S. M. Sapuan
Department of Mechanical and
  Manufacturing Engineering
Universiti Putra Malaysia
Serdang, Selangor, Malaysia

R. L. Sierakowski
AFRL
Eglin AFB, Florida, USA

W. J. Staszewski
Department of Mechanical
  Engineering
University of Sheffield
Sheffield, UK

I. N. Tansel
Department of Mechanical and
  Material Engineering
Florida International University
Miami, Florida, USA

K. Worden
Department of Mechanical
  Engineering
University of Sheffield
Sheffield, UK

P. S. M. M. Yusoff
Department of Mechanical
    Engineering
University of Technology Petronas
Tronoh, Perak Darul Ridzuan,
    Malaysia

1

Application of Artificial Neural Network in Composites Materials

M. Hasan, M. E. Hoque and S. M. Sapuan

CONTENTS

ABSTRACT An artificial neural network approach is a fascinating tool, inspired by the biological nervous system, that can be used to simulate a wide variety of complex scientific and engineering problems. A powerful artificial neural network function is determined largely by the interconnections between artificial neurons, similar to those occurring in their natural counterparts of biological systems. In composites, a certain amount of experimental results is required to train a well-defined neural network. Once the network has learned to solve the material problems, new data from the similar domain can be predicted without performing many long experiments.

1.1   Importance of Artificial Neural Network in Composites

Nowadays, the application of composites, especially polymer composites as engineering materials, has become state of the art. To be able to design the characteristics of composites is the most important advantage. To meet special engineering applications (e.g., concerning one or several measurable material properties), composites can be designed by selecting the correct composition and choosing the appropriate manufacturing processes. There are many possibilities to create composites containing different fillers and/ or reinforcements (Verpoest, 1998). In addition, not only different kinds but also various properties of matrix material and filler/reinforcements, as well as fiber concentration and, in case of continuous fiber composites, the laminate parameters can be considered. As a special example, one of the general principles for the systematic design of wear-resistant polymer composites is that the matrix should have high temperature resistance and high cohesive strength. Internal lubricants such as polytetrafluoroethylene and graphite flakes are frequently incorporated to reduce the adhesion and the frictional coefficient. Short aramid, glass, or carbon fibers are used to increase the creep resistance, hardness, and compressive strength of the polymer matrix. Additional fillers that enhance the thermal conductivity are also of great advantage (Friedrich et al., 2002).

Some key parameters in the manufacturing process, such as curing temperature and manufacturing speed, should also be analyzed. Property investigation plays a key role in materials science to evaluate composites designed for special engineering applications. All three stages are not separated but interconnected. The integration can be summarized as composite design, processing optimization, and property relationships. The first two fields correspond to the interaction between the selected compositions or the manufacturing process and properties investigated, whereas the last relates to possible correlations between some simple measured parameters and more complex properties. The understanding of all these relationships is important in composite material science to meet the requirements for particular engineering applications.

Modeling of the relationships generally involves the development of a mathematical tool derived from experimental data. Once established, it can significantly reduce the experimental work involved in designing new composites. This is why artificial neural networks (ANNs) have recently been introduced in the field of composites. Inspired by the biological nervous system, ANNs can be used to solve a wide variety of complex scientific and engineering problems. Like their biological counterparts, ANNs can learn from examples and can therefore be trained to find functional relationships without any prior assumptions about their nature. Furthermore, the network is built directly from experimental data by its self-organizing capabilities. Although there are few publications about the use of ANNs in the field of composites, the available literature at least covers various topics, from fatigue prediction to wear simulation and from monitoring of the manufacturing process to analysis of composite curing. It shows that the ANN approach can be properly employed in simulations of the interactions. It is therefore very important to summarize and recapitulate the developments and application of ANNs to promote wider consideration of the use of this potential mathematical tool in composite research.

1.2 Artificial Neural Network

Artificial neural networks are computational systems that simulate the microstructures of a biological nervous system. The most basic components of ANNs are modeled after the structure of the brain. Even the terminology is borrowed from neuroscience. The most basic element of the human brain is a specific type of cell, which provides us with the ability to remember, think, and apply previous experience to our every action. These cells are also known as neurons and each of these neurons can connect up to 200,000 other neurons. The power of the brain comes from the number of these basic components and multiple connections between them. All natural neurons have four basic components: dendrites, soma, axons, and synapses. Basically, a biological neuron receives input from other sources, combines them in some way, performs a generally nonlinear operation on the result, and then outputs the final result. Biological neural networks are constructed in a three-dimensional way from microscopic components. These biological neurons seem capable of almost unlimited interconnections (Aleksander and Morton, 1990; Fausett, 1994; Haykin, 1998; Bishop, 1995; Swingler, 1996).

Inspired by biological neurons, ANNs are composed of simple elements operating in parallel, that is, ANNs are the simple clustering of primitive artificial neurons. This clustering occurs by creating layers, which are then connected to one another. The way the layers connect may also vary. Basically, all ANNs have a similar topological structure. Some of the neurons interface with the real world to receive its input and other neurons provide the world with the network's output. All the remaining neurons are hidden from view. As in nature, the network function is determined by the interconnection between the neurons. Each input to a neuron has a weight factor of the function that determines the strength of the interconnections and thus the contribution of that interconnection to the following neurons. ANNs can be trained to perform a particular function by adjusting the values of these weight factors between the neurons either from the information from outside the network or by the neurons themselves in response to the input. This is the key to the ability of ANNs to achieve learning and memory.

1.3   Training Process of Back-Propagation ANNs

The multilayered neural network is the most widely applied neural network, which has been used in most researches so far. A back-propagation algorithm can be used to train these multilayer feed-forward networks with differentiable transfer functions to perform function approximation, pattern association, and pattern classification. The term *back propagation* refers to the process by which derivatives of network error, with respect to network weights and biases, can be computed. The training of ANNs by back propagation involves three stages (Fausett, 1994): (i) the feed forward of the input training pattern, (ii) the calculation and back propagation of the associated error, and (iii) the adjustment of the weights. This process can be used with a number of different optimization strategies.

1.4   Overfitting Problem

Multilayer feed-forward neural networks are believed to be able to represent any functional relationship between input and output if there are enough neurons in the hidden layers. However, too many neurons in the hidden layers may cause another problem, the so-called overfitting. The error on the training set is driven to a very small value because of the powerful ANN learning process, but when new data are presented to the network, the error is large. The network has memorized the training examples, but it has not learned to generalize to new situations. Clearly, when overfitting occurs, the network will not generalize well. The ideal method for improving network generalization is to use a network that is just large enough to provide an adequate fit. The larger the network, the more complex the functions that network can create. Therefore, if a small enough network is used, then it will not have enough power to overfit the data. Unfortunately, it is difficult to know beforehand how large a network should be for a specific application. There are two methods for improving generalization: regularization and early stopping (Demuth and Beale, 2000).

1.5   Applications of ANN to Composites

1.5.1   Fatigue Life

Fatigue is one of the most complicated problems for fiber composites, and failure mechanisms are still not well understood. Extensive tests must be carried out because of the absence of a well-defined failure criterion that can be used to predict fatigue failure in composites. ANNs offer the possibility of developing models that will predict the behavior of composites without being linked to mechanistic arguments. They have been introduced recently by Lee et al. (1999), Aymerich and Serra (1998), Al-Assaf and El Kadi (2001), and El Kadi and Al-Assaf (2002) for the prediction of fatigue life.

1.5.2 Unidirectional Composites

Al-Assaf and El Kadi (2001) applied the ANN approach to predict the fatigue life of unidirectional glass fiber/epoxy composites. Unidirectional fiber composite specimens were prepared with angle orientations of 0°, 19°, 45°, 71°, and 90°. They cyclically tested under load control condition with R ratios of 0.5, 0, and −1. Only the R value, the maximum stress, and fiber orientation angle were used as the ANN input, and the output was the number of cycles to fatigue failure. The database contained 92 measured results. ANN was applied, which showed a very satisfactory predictive quality, with an RMSE of less than 20%. It sees that the fiber orientation in unidirectional composites plays a key role in fatigue performance. Applying fiber orientation as an output improves the ANN predictive quality significantly, even with the 400 mentioned by Lee et al. (1999) for the laminates tested.

1.5.3 Laminate Composites

Lee et al. (1999) carried out an ANN prediction on the fatigue life of some carbon/glass fiber-reinforced plastic laminates. As the first step, an extensive existing database of fatigue lives for four common glass fiber-reinforced plastics—HTA/913, T800/5245, T800/924, and IM7/977—was used to evaluate possible ANN architectures and to develop training methods for the network selected for detailed study. The database contained more than 400 fatigue results over R ratios of −1.5, −1.0, −0.3, +0.1, and +10. Three fatigue parameters (peak stress, minimum stress, probability of failure) and four monotonic mechanical properties (tensile strength, compressive strength, tensile failure strain, and tensile modulus) were selected as the ANN input, which are applied to predict the fatigue life of the composites as the outputs. ANN was finally optimized by evaluating the changes in RMSE of network output with the number of neurons in the hidden layer. Further effort was carried out to reduce the size of training database and satisfactory prediction was achieved. It was their clear conclusion that the ANNs (a) can be trained at least to model constant-stress fatigue behavior as well as other current life-prediction methods and (b) can provide accurate presentations of the stress/R ratios/median-life surfaces for carbon fiber composites from a data set of quite small experiments.

1.5.4 Brief Comments

To conclude, successes have been achieved using ANN in the prediction of fatigue life of composites. However, fatigue behavior is still so complicated that the problem requires more effort before the ANN can be used with more confidence. Other publications on fatigue prediction (Venkatesh and Rack, 1999; Haque and Sudhakar, 2001; Pleune and Chopra, 2000) in the field of composites are also recommended for the improvement of this technique.

1.6 Tribological Properties

1.6.1 Wear of Composites

Velten et al. (2000) were one of the earliest pioneers to explore the ANN approach in composites to predict the wear volume of short-fiber/particle-reinforced thermoplastics. A total data set of 72 independent wear measurements was used to train and test the neural network. The data set came from fretting tests with various material compositions under different wear-measuring conditions. ANN was applied with the output of wear volume. The inputs were mechanical properties and test conditions that include compressive strength, compressive modulus, compressive strain to failure, tensile strength, tensile strain to failure, impact strength, environmental testing temperature, initial load, average load, and average velocity. An automated "Bayesian" regularization of a back-propagation algorithm was selected, which has the capacity of automatically identifying the optimal size of the ANN in its hidden layer. A randomly chosen test data set was used in the quality evaluation. Some success was achieved in this attempt at property analysis using ANNs to deal with wear of composites.

In further work, Zhang et al. (2002) carried out improvements based on an enlarged data set of 103 independent measurements. The database contained (1) the material composition, (2) mechanical properties of composites, and (3) testing conditions as the input parameters (Zhang and Friedrich, 2002). The wear characteristics such as specific wear rate or frictional coefficient were chosen as the output data. In all the tests, 88 measurements were used for training and the remaining 15 for testing. Again, Bayesian algorithm was applied. The ANN obtained a good predictive quality of 38% test data.

1.6.2 Erosive Wear of Polymers

Recently, Zhang et al. (2003) applied ANN to deal with erosive wear data for three polymers: polyethylene, polyurethane, and epoxy. The impact angle for solid particle erosion and some characteristic properties was selected as

ANN input variables for predicting the erosion wear rate. To investigate the correlations between erosive wear rate and the characteristic properties of these polymers, each characteristic property was used only with the necessary erosion conditions as input variables for training the ANN. The qualities were analyzed by the percentage, which was used to rank the importance of these characteristic properties. The ANN obtained a good predictive quality.

1.7 Dynamic Mechanical Properties

Zhang et al. (2002) provided an ANN prediction based on the measurement results of dynamic mechanical properties of short fiber-reinforced composites in a temperature range from −50°C to 150°C. Material compositions and various measuring temperatures were selected as the ANN input parameters and linear viscoelastic properties acted as output variables. Twenty-five neurons were arranged in the hidden layer. Bayesian regularization algorithm was selected, which can automatically identify the optimal size and structure of the hidden layer. The total number of data were 480. The data set was divided into a training data set and a test data set. The number of training data was firstly investigated because it strongly influences the predictive quality. Again, the ANN obtained a good predictive quality.

1.8 Processing Optimization

A neural network has been used by Allan et al. (2001) to predict the structure and properties of polypropylene fibers, which could be helpful in identifying the control parameters in manufacturing. Key manufacturing parameters, such as temperature and speed of the rollers and plates, were selected as the ANN inputs. The outputs were fiber properties, tenacity, elongation, modulus, shrinkage, crystal order, and orientation, which mainly concern the quality of the fibers. The predictions were based on multilayered ANN and several encouraging results were obtained, which showed that the ANNs were able to generalize quite well in the simulation of manufacturing processes.

1.9 Other Related Applications

Conventional optimization methods for composite structures, on one hand, usually involve the development of step-by-step procedures and their implementation into a larger computer program. The application of ANN approach offers, on the other hand, a new route in which a large number of simple processing units are directly connected within the network structure by means of weights. An ANN has the capacity to learn the underlying principles involved in solving the optimization problem from examples by adjusting these weight factors. After a well-trained ANN is obtained, it then can be used to tackle similar problems in optimal design of composite structures.

1.10   Conclusions

Evolving from neurobiological insights, the neural network approach gives a computer system an amazing capacity to actually learn from input data. An ANN is ideally suited for simulating complex composite problems because, like its biological counterparts, it can learn and therefore can be trained to find solutions. The following conclusions can be drawn from the current review:

i. Composite design. It is ideal in composite design where only material compositions and testing conditions serve as ANN input data. A well-trained ANN is expected to be very helpful in predicting the material properties before manufacturing/testing the real composites.

ii. Processing optimization. There are many parameters in the manufacturing processes that control the final qualities of the composites. ANNs could be of help to simulate the relationships between these manufacturing parameters and material performance, which can be used as a basis for a computer-based processing optimization.

iii. Property analysis. The analysis of relationships between some simple properties and other complex properties will be of additional help in the design of new composite materials. The simple properties are normally easier to obtain than complex ones and, therefore, successful prediction could be of benefit to reduce the number of more complex experiments.

iv. Size of the training data set. The required number of training data can be reduced by optimizing the neural network architecture and by choosing suitable input parameters. However, the more complex the nonlinear relation between input and output is, the more training data are required.

References

Al-Assaf, Y., and El Kadi, H. 2001. Fatigue life prediction of unidirectional glass fiber/epoxy composite laminate using neural networks. *Composite Structures* 53:65–71.

Aleksander, I., and Morton, H. 1990. *An introduction to neural computing*. London: Chapman & Hall.

Allan, G., Yang, R., Fotheringham, A., and Mather, R. 2001. Neural modelling of polypropylene fiber processing: predicting the structure and properties and identifying the control parameters for specified fibers. *Journal of Materials Science* 36:3113–3118.

Aymerich, F., and Serra, M. 1998. Prediction of fatigue strength of composite laminates by means of neural networks. *Key Engineering Materials* 144:231–240.

Bishop, C. M. 1995. *Neural networks for pattern recognition*. Oxford: Oxford University Press.

Demuth, H., and Beale, M. 2000. Neural network toolbox users guide for use with MATLAB version 4.0. Math Works.

El Kadi, H., and Al-Assaf, Y. 2002. Prediction of fatigue life of unidirectional glass fiber/epoxy composite laminate using different neural network paradigms. *Composite Structures* 55:239–246.

Fausett, L. V. 1994. *Fundamentals of neural networks*. 1st ed. Englewood Cliffs, NJ: Prentice Hall.

Friedrich, K., Reinicke, R., and Zhang, Z. 2002. Wear of polymer composites. *Journal Engineering Tribology, Proceedings of the Institution of Mechanical Engineers* 216: 415–426.

Haque, M., and Sudhakar, K. V. 2001. Prediction of corrosion-fatigue behavior of DP steel through artificial neural network. *International Journal of Fatigue* 23:1–4.

Haykin, S. S. 1998. *Neural networks: a comprehensive foundations*. 2nd ed. Upper Saddle River, NJ: Prentice Hall.

Lee, J. A., Almond, D. P., and Harris, B. 1999. The use of neural networks for the prediction of fatigue lives of composite materials. *Composites Part A: Applied Science and Manufacturing* 30:1159–1169.

Pleune, T. T., and Chopra, O. K. 2000. Using artificial neural networks to predict the fatigue life of carbon and low alloy steels. *Nuclear Engineering and Design* 197:1–12.

Swingler, K. 1996. *Applying neural networks: a practical guide*. London: Academic Press.

Velten, K., Reinicke, R., and Friedrich, K. 2000. Wear volume prediction with artificial neural networks. *Tribology International* 33:731–736.

Venkatesh, V., and Rack, H. J. 1999. A neural network approach to elevated temperature creep-fatigue life prediction. *International Journal of Fatigue* 21:225–234.

Verpoest, I. personal communication, 1998.

Zhang, Z., and Friedrich, K. 2002. Artificial neural network in polymer composites. Proceedings of the 3rd Asian-Australasian Conference on Composite Materials 105–118.

Zhang, Z., Barkoula, N.-M., Karger-Kocsis, J., and Friedrich, K. 2003. Artificial neural network predictions on erosive wear of polymers. *Wear* 255:708–713.

Zhang, Z., Friedrich, K., and Velten, K. 2002. Prediction of tribological properties of short fiber composites using artificial neural networks. *Wear* 252:668–675.

Zhang, Z., Klein, P., and Friedrich, K. 2002. Dynamic mechanical properties of PTFE based short carbon fiber reinforced composites: experiment and artificial neural network prediction. *Composite Science and Technology* 62:1001–1009.

2

Neural Network Approaches for Defect Detection in Composite Materials

T. D'Orazio, M. Leo, and C. Guaragnella

CONTENTS

ABSTRACT The problem of detecting internal defects in composite materials is felt to be unavoidable in many industrial contexts both for quality control of production lines and for maintenance operations during in-service inspections. Among nondestructive techniques, thermographic image analysis has received much attention for the inspection of composite materials. In this chapter, we address the problem of developing neural network approaches for an automatic system of defect detection by the analysis of sequences of thermographic images. Neural networks are very promising because they offer the opportunity to associate input signals to output classes even in the case of nonlinear mapping. In particular, we have considered supervised and unsupervised approaches for training neural network and we have discussed the pros and cons of their applicability in systems that could help safety inspectors in the diagnosis of problems.

2.1 Introduction

The problem of guaranteeing reliable and efficient safety checks has received much attention in recent years in many industrial contexts. Quality control and maintenance operations not only have to be reliable but also have to be performed at low cost to frequently meet schedules. In particular, non-destructive testing and evaluation techniques are necessary to detect damage in high-stressed and fatigue-loaded regions of the structure at an early stage. Some of these nondestructive testing and evaluation techniques are based on the analysis of different signals such as ultrasonics, acoustic emission, thermography, laser ultrasonic, X-radiography, eddy currents, shearography, and low-frequency methods (Huang et al., 2001). Transient thermography is a very promising technique for the analysis of composite materials (Gaussorgues, 1994). It is a noncontact technique, which uses the thermal gradient variation to inspect the internal properties of the investigated area. The materials are heated by an external source (lamps) and the resulting thermal transient is recorded using an infrared camera. Some research has been presented in literature on the use of thermography (Jones, 1995; Avdelidis et al., 2003; Wu and Busse, 1998). They have demonstrated the effectiveness of thermography in detecting internal defects and show excellent results on all the investigated samples. Different qualitative approaches have been developed by many researchers to investigate the effects on thermographic images of a number of parameters such as specimens of materials; defect types; and depths of defect, size, and thickness (Sakagami and Kubo, 2002; Giorleo et al., 2000; Inagaki et al., 1999). Quantitative approaches are attractive in the analysis of thermographic images because of the possible diagnosis capabilities that they introduce. They involve the solution of the direct problem (i.e., the computation of the expected response from known sound and defective materials) and the inverse problem (i.e., the evaluation of defect characteristics from a known response). Because of the nonlinear and nonunivocal nature of these mapping problems, the solution is rather complex. For this reason, some attempts using neural networks have started to emerge in the last few years (Maldague et al., 1998; Marin and Tretout, 1999; Saintey and Almond, 1997; Vallerand and Maldague, 2000; Galietti et al., 2007; Just-Agosto et al., 2008).

In this chapter, we address the problem of developing an automatic system for the analysis of sequences of thermographic images to help safety inspectors in the diagnosis of problems. Starting from the observation that composite materials have different behaviors during the transient phase of the thermographic inspection and that the reflectivity time evolution of each pixel of the image differs when some regions contain inner defects, we devised neural approaches that analyze the main characteristics of these thermic evolutions, extract significant information, and then use them to classify the investigated area as a defective or sound area.

We tackle first the problem of proper preprocessing required for extracting features significant to the defect recognition phase. Preprocessing a thermographic transient response of a composite material needs—at first—a noise reduction stage. Space domain filtering, often used, introduces a blurring effect reducing the image resolution of the obtained video sequence. Because of the physics of the thermal transmission through materials, neighboring pixels have a time-correlated thermal evolution so that their information content can be used to reduce the noise, reducing the image blurring effects and enhancing segmentation of the processed image. With this aim, we compare time denoising techniques with time-space denoising techniques before the application of neural classifiers. Then, we address the problem of signal classifications using automatic learning paradigms. Neural networks are particularly suitable for these problems because they have the natural propensity for storing exponential knowledge and making it available for use. We compare and discuss two different learning approaches: supervised and unsupervised learnings. They differ in the quantity of information that the environment has to provide to train the networks. Supervised approaches require a set of input–output examples, whereas unsupervised approaches require a measure of the similarity that is acceptable for data aggregation. We have compared the results of the application of different neural classifiers and different preprocessing techniques. Experiments on a composite material containing several defects demonstrated the effectiveness and the potential capabilities of the proposed approaches.

2.2   Signal Processing

The thermographic analysis produces a sequence of images in which the value of each pixel (i,j) represents the temperature variation during the heating and warming phases.

In Figure 2.1, the monodimensional signals extracted from the thermographic sequence of some points belonging to sound and defect areas are plotted. The points were selected from different regions belonging to different kinds of defects. From the graph it is clearly evident that a functional description of the intensity variations cannot be easily generalized and the behaviors of points corresponding to different defective areas are very similar. A preprocessing technique is required both for noise reduction and for extraction of significant information that could increase the probability of separating signals belonging to different areas. It consists of two different preprocessing techniques: the temporal signals were filtered applying the wavelet transform and considering different filter bank configurations. Then, it analyzed the spatiotemporal variations of the thermographic signals

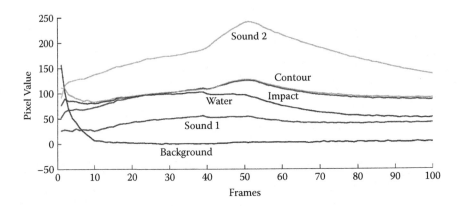

FIGURE 2.1
Some monodimensional signals extracted from the temporal sequences of thermographic images.

as have been observed that neighbor pixels have related temperature variations that can be used for a denoising procedure on the original signals.

2.2.1 The Wavelet Filtering

Wavelet transform is an extension of the Fourier transform that allows better representation in the case of nonstationary signals. In fact, Fourier representation points out whether a certain frequency component exists or not in the considered signal, but this information is independent of where in time this component appears. Wavelet representation instead represents signal contents at different frequencies with different resolutions. In particular, wavelet representation is designed to give good time resolution and poor frequency resolution at high frequencies and good frequency resolution and poor time resolution at low frequencies. Let $f \in L^2(R)$ be an $L^2(R)$ function, the wavelet transform of f at scale s is defined as

$$F^W(u,s) = \frac{1}{\sqrt{s}} \int_{-\infty}^{+\infty} f(t)\psi\left(\frac{t-u}{s}\right) dt$$

where the base atom ψ is a zero average function, centered around zero with a finite energy. The modulation of the function ψ by the factor $1/s$ increases its amplitude when the scale s decreases and vice versa. In terms of frequencies, we can state: for small scales s, ψ_s has high frequencies; and as s increases, the frequency of ψ_s decreases. In this way, a multiresolution representation of the function f can be achieved. Notice that the wavelet transform can be written as a convolution product $F^W(u, s) = f * \psi_s(u)$ (it is a linear space-invariant filter).

This leads to a fast and efficient implementation of the wavelet transform for discrete signal obtained using digital filtering techniques. The signal to be analyzed is passed through filters with different cutoff frequencies at different scales. The wavelet transform for discrete signal is computed by successive low-pass and high-pass filtering of the discrete time-domain signal. Many filter kernels can be used for this scope and the best choice depends on the features of the input signal that have to be exploited.

At each decomposition level, the half-band filters produce signals spanning only half the frequency band. This doubles the frequency resolution as the uncertainty in frequency is reduced by half. At the same time, the decimation by 2 doubles the scale. With this approach, the time resolution becomes arbitrarily good at high frequencies, whereas the frequency resolution becomes arbitrarily good at low frequencies.

In Figure 2.2, a three-level wavelet decomposition is shown. The signal $x(n)$ is decomposed and its final representation consists of approximation coefficients at level 3, $A_n(3)$, and detail coefficients $D_n(1)$, $D_n(2)$, and $D_n(3)$.

In this work, each monodimensional signal extracted from the thermographic sequence at position (i,j) is given as input to a filter bank. In particular, two kinds of filter kernels have been used, namely, Daubechies and biorthogonal. Daubechies are widely used for many application problems because of their kernel orthogonality that is a very desirable property because it allows us to project data onto orthonormal spaces and then to better enhance signal characteristics for further processing. There exists many Daubechies filter types: type 1 filters (also referred as Haar or DB1) are the most used because they also perform symmetric analysis and can be implemented in a very fast way (this makes them well suited for real-time applications). In Figure 2.3, some Daubechies wavelet bases are reported.

Biorthogonal wavelets constitute a generalization of orthogonal wavelets: instead of a single orthogonal basis, a pair of dual biorthogonal basis functions is employed (Figure 2.4). The pair of functions is used to share the workload: one function of the pair acts as the analyzing function, whereas the other acts as the reconstruction function. That the roles of these functions can be interchanged is called *duality principle*. The basis function is then defined recursively by two pairs of filters. Designing biorthogonal wavelets allows

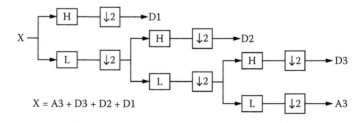

FIGURE 2.2
Three-level wavelet decomposition.

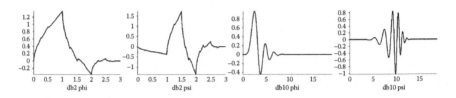

FIGURE 2.3
Some Daubechies wavelet bases.

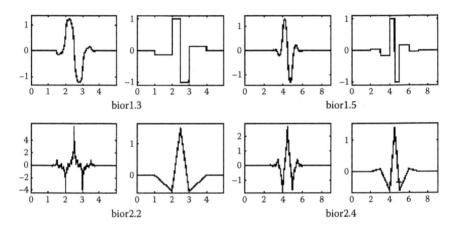

FIGURE 2.4
Some biorthogonal wavelet bases.

more degrees of freedoms than orthogonal wavelets. One additional degree of freedom is the possibility to construct symmetric wavelet functions: in many applications, the symmetry of the filter coefficients is often desirable because it results in linear phase of the transfer function. In Figure 2.3, some Daubechies wavelet bases are reported. The result of the decomposition at level 4 of a thermographic signal is reported in Figure 2.5.

2.2.2 The Single Value Decomposition SVD Preprocessing

For each point (i,j) of the thermographic sequence, the 3×3 window obtained considering the eight neighboring pixels has been extracted along the time as shown in Figure 2.6. Observing these extracted signal sets, two things appeared evident: the signal evolution is very slow and the time evolutions of neighbor pixels are similar (see Figure 2.7). For this reason we decided to exploit all the time correlation coming from signals to reduce as much as possible the unwanted noise. The singular value decomposition (SVD) is applied to this $N \times 9$ matrix (where N is the length of the thermographic sequence) and the more significant components are extracted and used to

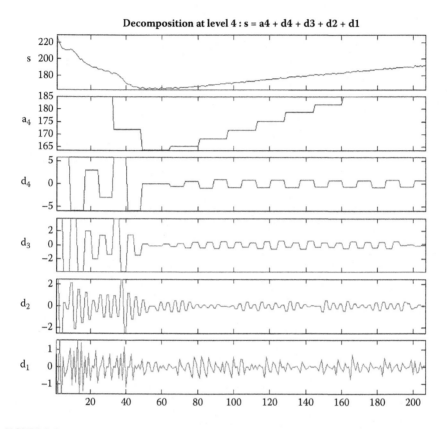

Decomposition at level 4 : s = a4 + d4 + d3 + d2 + d1

FIGURE 2.5

Decomposition at level 4 of a thermographic signal. The original signal is on top. Then, approximation and detail coefficients are shown.

reconstruct the signal *(i,j)*. In this way, we provided a good approximation of the original signal but with reduced noise.

Let M be the data $N \times 9$ matrix containing in the columns the time evolution of the signals, the SVD decomposition is defined in a matrix form as:

$$M = U \cdot S \cdot V^{H} \tag{2.1}$$

where V^{H} represents the transpose of the matrix V.

The matrices M and U are $N \times 9$, whereas V is a 9×9 orthonormal matrix (the eigenvector matrix). Matrix S is diagonal, containing the signals in U modula, of a vector A, defined as

$$A = M \cdot V = U \cdot S \tag{2.2}$$

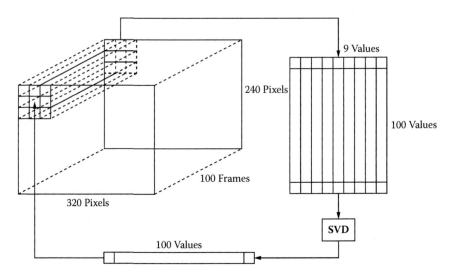

FIGURE 2.6
The 3 × 3 × 100 window extracted from the temporal sequences of thermographic images on which the SVD is applied.

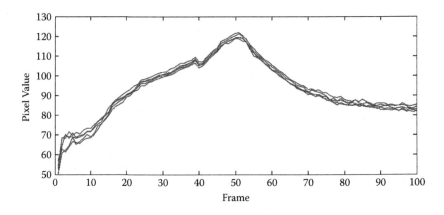

FIGURE 2.7
The time evolutions of neighboring points belonging to a sound area.

related to the squared mean values of the orthogonal signals contained in

the matrix A, but for a scaling factor of $\sqrt{\dfrac{1}{N}}$:

$$S_{i,i} = \sqrt{\sum_{J=1}^{N} A_{j,i}^2}$$

(2.3)

The eigenvectors in the matrix V are coefficients of the "filters" used to select the correlated components of the signals in the matrix M.

The first singular value is very high, whereas all the others very often correspond to noise contribution, here defined as the uncorrelated signal components of the data matrix M. The decomposed image can be divided into two submatrices, the signal matrix (M_o) and the noise matrix (M_n). Such matrices can be obtained from the decomposition by recombining the original matrix using the singular value matrix S_o, obtained by assigning zero to all eigenvalues but the first in S and $S_n = S - S_o$.

$$M_o = U \cdot S_o \cdot V^H \qquad (2.4)$$

and

$$M_n = U \cdot S_n \cdot V^H \qquad (2.5)$$

The largest part of the energy content of the correlated information in the vector observation is contained in the first column of the matrix U, in this way preserving the time signal shape. Furthermore, as the first orthogonal vector of A in (2.2) is obtained as a weighted sum of highly correlated vector components, the signal-to-noise ratio increases with the number of components in the observation vector. In our case, as we process nine pixels of a 3 × 3 pixel neighborhood, the maximum increase in the signal-to-noise ratio would be 9 (i.e., about 10 dB). The signals coming from the image reconstruction with the S_o matrix are all similar to each other, so that only the signal corresponding to the central pixel is retained; a new video sequence is constructed applying the same procedure to all the 3 × 3 pixel neighborhoods of the image on the original video; the proposed denoising procedure is used to construct the output denoised video using the principal component of the pixel time evolution corresponding to the center pixel reconstructed in Equation (2.4). The proposed procedure is able to guarantee some advantages with respect to standard filtering procedure:

- The filtering procedure does not use heuristic filters, but data-adaptive ones, able to exploit all the correlation inherently present in the thermal time evolution signals;
- The adaptivity of the denoising procedure allows blind filtering, that is, the user does not have to deal with the defect type or characteristics, the material type, or the heating sources used;
- The blurring effects coming from the spatial filtering procedure are reduced as the filtering effect takes place in the time domain, but results of the filtering procedure enhance the space domain (image quality).

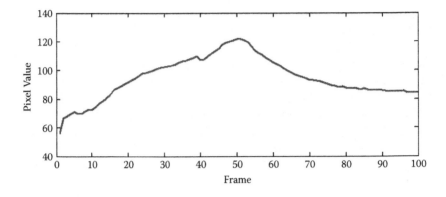

FIGURE 2.8
The time evolution of the central point reconstructed with the first component of the SVD evaluated in its neighboring points (3 × 3) in Figure 2.7.

In Figure 2.8, the signal reconstructed with the first component of the SVD evaluated in its neighboring points is reported. Comparing this signal with the initial ones in Figure 2.7, a noise reduction especially in the initial and final parts of their time evolutions is clearly visible.

2.3 Neural Classifiers

In the last decade, neural networks have been largely applied to pattern recognition problems, first, for their massively parallel structure and, second, for their ability to learn and therefore generalize (Haykin, 1994; Freeman and Skapura, 1991). Generalization is very important for classification tasks because it refers to the capability of producing reasonable outputs for inputs not encountered during training. Neural networks learn through an iterative process of adjustments of the internal weights and thresholds. Ideally, the network becomes more knowledgeable about the task after several iterations of the learning process. Learning algorithms differ from each other in the manner in which a neural network relates to its environments. Supervised or active learning paradigms require the a priori knowledge of a set of input–output examples; unsupervised learning paradigms need a task independent measure of the quality representation that the network is required to learn. The problem of the automatic classification of thermographic signals corresponding to defective or sound areas can be considered either as a supervised learning problem or as an unsupervised learning one. In many cases, especially during well-known production processes, the possible defects detectable in composite materials are predictable. Sample materials with many defect areas are available and used to verify the good functioning of the inspection sensors as an initialization procedure before every test. In this case, some signals corresponding to defect areas and sound areas can be extracted and used as sets of input–output examples for training the neural network. In many other contexts, such as during in-service inspection, it is not possible to know the defects that can be found in composite materials. The only information available can be the general uniformity of signals corresponding to sound areas. When something differs from this uniform behavior, it can be suspected as belonging to a defective area. In these cases, unsupervised approaches are well suited for signal classification.

In the following subsections, we describe two different classification methods: a multilayer neural network trained with a back-propagation algorithm for the supervised learning problem and a self-organizing map that uses an unsupervised algorithm to aggregate the input data.

2.3.1 Multilayer Feed-Forward Networks

Multilayer feed-forward networks have been applied successfully to solve some difficult and diverse problems by training them in a supervised manner. These networks consist of a set of sensory units that constitute the input layer, one or more hidden layers of computation nodes, and an output layer of computation nodes. The input signal propagates through the network in a forward direction on a layer-by-layer basis (see Figures 2.9 and 2.10).

Each computational node (referred to as a neuron) is an information unit that consists of three basic elements: a set of synapses (connection links with other nodes), each of which is characterized by a weight that establishes the node excitation/inhibition level in correspondence to each input, a linear

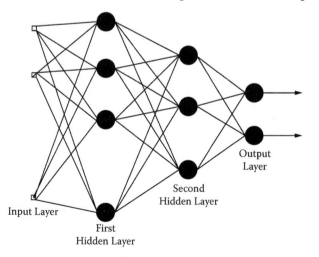

FIGURE 2.9
A multilayer perception with two hidden layers.

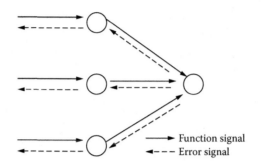

Function signal
Error signal

FIGURE 2.10
A multilayer perception with two hidden layers.

combiner that sums up the input signals weighted by the respective synapses, and an activation function that determines the amplitude of the output node.

The activation functions can be linear, piecewise linear, and nonlinear according to the computational capability required by each node. In the same way, different network architectures can be devised according to the complexity of the classification problem (input/output transformation not separable by a linear function). The presence of hidden neurons enable the network to learn complex tasks by progressively extracting more meaningful features from the input patterns.

These networks are trained with a popular algorithm known as error back-propagation algorithm. This algorithm is based on the error-correction learning rule. The learning process consists of two phases: a forward pass and a backward pass. In the forward pass, the input vector is applied to the sensory nodes of the network and its effect propagates through the network layer by layer until a set of outputs is produced. During the forward pass, the synaptic weights of the network are fixed. During the backward pass, the synaptic weights are all modified in accordance with the error-correction rule. An error signal is produced proportional to the difference between the desired response of the network and the actual response. This error signal is propagated backward through the network so as to make the actual response of the network more closer to the desired response.

The number of input nodes depends on the length of input vectors. In the case of continuous signals, the number of input nodes is the number of points obtained by the signal uniform sampling. On the other side, the number of nodes in the output layer depends on the number of classes required by the problem under consideration. The most difficult task that has to be solved during the designing of a neural classifier is the definition of the number of hidden layers with the corresponding number of nodes. The last point has to be solved with different trials and after the evaluation of the generalization capability of the resulting neural networks.

2.3.2 Self-Organizing Map Neural Network

A self-organizing map neural network (commonly referred as SOM) is a winner-take-all artificial neural network that discovers similarity in the data and builds a bidimensional map of M neurons organized in such a way that similar data are mapped onto the same neuron or to neighboring nodes in the map. This leads to a spatial clustering of similar input patterns in neighboring parts of the SOM, and the clusters that appear on the map are themselves organized. This arrangement of the clusters on the map reflects the attribute relationships of the clusters in the input space. At first, each neuron i of the map is assigned an n-dimensional weight vector, where n is the dimension of the input data. The training process of SOMs may be described in terms of data presentation and weight vector adaptation. Each training iteration t starts with the random selection of one input vector x. This vector is presented to the SOM and each neuron of the net determines its activation. The Euclidean distance between the weight vector and the instance is used to calculate a unit's activation. The unit with the lowest distance is then referred to as the winner, c. Finally, the weight vector of the winner as well as the weight vectors of selected units in the vicinity of the winner is adapted. For the units in the vicinity of the winner, a gradual adaptation is performed: the lower the distance, the greater the updating. The updating formula for a neuron with weight vector $W(t)$ can be expressed as

$$W(t+1) = W(t) + \Theta(v,t)\alpha(t)(x(t) - W(t))$$

where $\alpha(t)$ is a monotonically decreasing learning coefficient, $x(t)$ is the input vector, and $\Theta(v,t)$ is the neighborhood function depending on the lattice distance between the input and the neuron v. The weight vectors of the adapted units are moved slightly toward the neurons. The amount of weight vector movement is guided by the learning rate, which decreases over time. The number of units that are affected by adaptation as well as the strength of adaptation depending on a unit's distance from the winner is determined by the neighborhood function. This number of units also decreases over time such that toward the end of the training process, only the winner is adapted.

The neighborhood function is unimodal, symmetric, and monotonically decreasing with increasing distance to the winner, for example, Gaussian. The movement of weight vectors has the consequence that the distance between the inputs and weight vectors decreases. In other words, the weight vectors become more similar to the input. Hence, the respective unit is more likely to win at future presentations of this input. At the end of the training process, each input activates a neuron of the map and similar input activates adjacent neurons on the bidimensional map. The resulting SOM allows us to determine, for each input vector, the relative neuron into the map, but to perform online classification of unknown input instances, a further step is necessary. Considering that usually the number of neurons in the map is

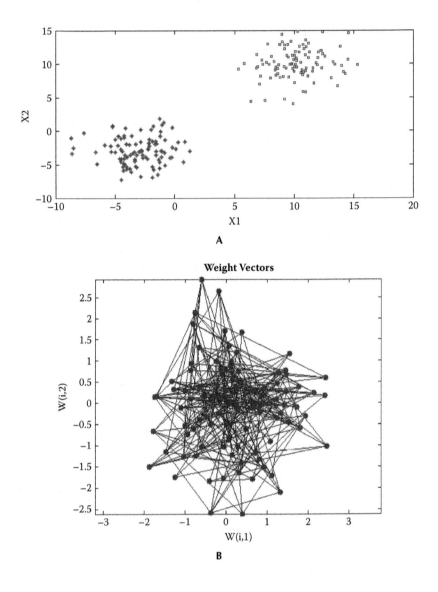

FIGURE 2.11
(A) The data that have to be clustered by the SOM. (B) The initial weights of the SOM.

lower than the number of the input instances but is greater than the number of classes of the instances, it is necessary to group (cluster) neighboring neurons. The number of groups has to equal the number of classes to be determined in the data. In Figure 2.11, an example of data aggregation by the SOM is shown. Two clusters of bidimensional data are provided to the SOM (Figure 2.11A). The initial random weights are shown in the image on the left (Figure 2.11B). After 1000 presentations of the input data, the weights move with a gradual adaptation phase (Figure 2.11C): it is possible to see the two aggregations of weights that can be clustered in a successive phase.

In this chapter, a K-means clustering of the neurons in the SOM has been performed, where the number K of clusters has been set to the number of classes to be determined. K-means is an unsupervised clustering algorithm that defines k centroids, one for each cluster to be determined. These centroids should be carefully placed because different location causes different result. They are initially placed on a regular grid and converge with the iterations to the centroids of the data observations. The next step is to take each point belonging to a given data set and associate it to the nearest centroid. When no point is pending, the first step is completed and an

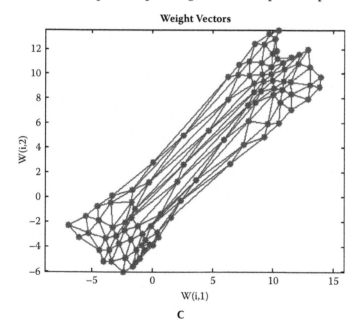

Weight Vectors

C

FIGURE 2.11C
The weights of the SOM after 1000 iterations of the learning phase. The adaptation phase moved the weights according to the input data.

earlygrouping is done. At this point we need to recalculate k new centroids as barycenters of the clusters resulting from the previous step. After we have these k new centroids, a new binding has to be done between the same data set points and the nearest new centroid. A loop has been generated. As a result of this loop, we may notice that the k centroids change their location step by step until no more changes are done. In other words, centroids do not move any more. Finally, this algorithm aims at minimizing an objective function:

$$ J = \sum_{j=1}^{k} \sum_{i=1}^{n} \left\| x_i^{(j)} - c_j \right\|^2 \tag{2.7} $$

where $\left\| x_i^{(j)} - c_j \right\|^2$ is a chosen distance measure between the n data points $x_i^{(i)}$ and the cluster center c_j.

2.4   Experimental Results

2.4.1   Experimental Setup

The thermographic image sequence was obtained by using a thermo camera sensitive to infrared emissions. A quasi-uniform heating has been used that guarantees a temperature variation of the composite materials around 20°C/ sec. Five images per second were acquired (100 images in the sequence), obtaining for each pixel a thermographic signal of 100 points. The composite material used in the experimental tests has an alloy core with a periodic honeycomb internal structure. It presents different kinds of defects: specifically, there is a hole with some water and five impact damages. In all the cases, the defects or the internal damage are not detectable with a visual inspection. In Figure 2.12, one of the thermographic images is reported. The defect type classification is superimposed on the image. A number of experiments were carried out.

In the first experiment, the original thermographic images were segmented without preprocessing: thermographic signals were provided as input both to a neural network classifier trained with a back-propagation algorithm (referred for brevity as BP classifier) and to a self-organizing map neural network (referred as SOM classifier). To train the BP classifier, a number of signal samples for each class (50 examples for impact damages, 50 examples for water seepage, and 100 examples for nondefective area) was manually selected from the images and provided to the neural network that, gradually, became able to recognize defective areas and segment them from the background (D'Orazio et al., 2005). The same training set was used in all the experiments that will be described later with the BP classifier. The neural

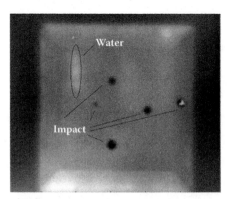

FIGURE 2.12
One of the thermographic images.

network consisted of 100 input nodes (the number of points of the original thermographic signals) and three output nodes (the number of classes that we want to recognize). When the value of the output node was greater than a fixed threshold (0.65), the node was considered as the winner and the signal was associated to the corresponding class. We decided also to consider a range of uncertainty (0.4–0.65) in which we cannot associate the signal to any class. These points were found on the border regions of defect areas. The results are reported in Figure 2.13. The segmented image obtained with the BP classifier and the segmentation result obtained with the SOM classifier are shown, respectively, at (A) and on (B). Different colors were used to represent different classes. In the image on the left, the contour regions of defective areas have different colors because they belong to the uncertain class (actually in this image, four colors were used because of the impact, water, and sound classes and the uncertain class). Notice that both images are affected by noise and the defect areas are not clearly visible. In the following subsections, the results obtained introducing wavelet and SVD preprocessing in combination with BP and SOM classifiers are reported and the advantages of their use are clearly pointed out.

2.4.2   Wavelet Preprocessing with Neural Networks

In this section the experiments performed by using wavelet preprocessing are discussed. Wavelet coefficients have been given as input to BP and SOM classifiers. Different filter kernels (Daubechies, biorthogonal, etc.), decomposition levels, coefficient selection approaches, and network configurations (number of hidden neurons, kind of activation functions) were tested to find the best defect detection solutions.

The segmentation results with BP classifiers were generally satisfying. In Figure 2.14, two experimental outcomes using approximation coefficients at different decomposition levels after Daubechies (type 1) filtering and 20 hidden

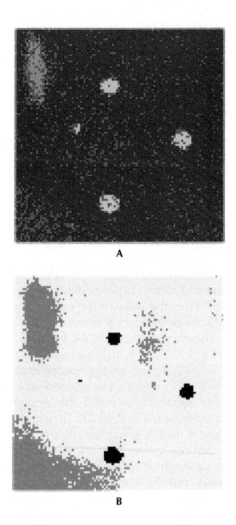

FIGURE 2.13
Results obtained without any preprocessing. (A) The segmented image with the BP. (B) The segmented image with the SOM. Different colors represent different classes.

neurons with sigmoidal activation function are reported. In particular, in the image on the left, the wavelet approximation coefficients at level 1 were used (generating signals of 50 points), whereas in the image on the right, the wavelet approximation coefficients at level 5 (generating signals of 4 points) were given as input to the BP classifier. It should be observed that in the image on the right, the segmented defective areas contained some holes; their edges were not well defined; and, on the bottom left side, a false defective region persisted due to lighting effects. The image on the left was the best segmentation result: only real defective areas were segmented and their edges were finely localized.

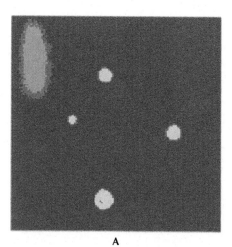

A

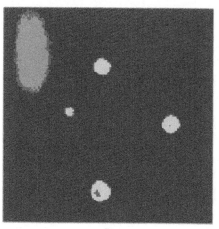

B

FIGURE 2.14
Results obtained with the wavelet preprocessing and BP classifiers. (A) The results with the walevet approximation coefficients at level 1. (B) The results with wavelet approximation coefficients at level 5.

In the experiments performed by using the SOM classifier, the wavelet coefficients were classified without any preliminary training phase. Input vectors were, in fact, projected on a map and then grouped by the well-known K-means clustering method. In this way the segmentation results were independent of the selection of training samples from defect and sound areas. As a drawback, the segmentation outcomes were less accurate. In Figure 2.15, two experimental outcomes are reported. In particular, (A), the best defect segmentation results obtained using a SOM classifier is reported: the input of

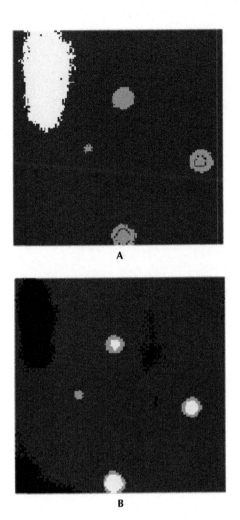

FIGURE 2.15
Results obtained with the wavelet preprocessing and SOM classifiers. (A) The results with the walevet approximation and detail coefficients at level 1. (B) The results with wavelet approximation coefficients at level 3.

the SOM (with a 10 × 10 projection map) were both wavelet approximation and detail coefficients at level 1 obtained by Daubechies (type 1) kernels. On the right, another experimental result is shown: in this case, the wavelet approximation coefficients at level 3 (biorthogonal kernel) were given as input to the SOM classifier (with a 10 × 10 projection map). The results can be considered satisfying even if the defect regions were not well segmented as in the previous experiment. The SOM classifier was able to separate the two defective areas from the sound areas by only considering the signal similarity.

2.4.3   SVD Preprocessing with Neural Networks

This section describes the experiments performed by applying the SVD to decompose the original thermographic image sequence and then using the first components to reconstruct the images.

Figure 2.16 shows the results obtained by the back-propagation neural classifier applied to the thermographic images after the reconstruction with the components S_1 (Figure 2.16A) and S_1S_2 (Figure 2.16B).

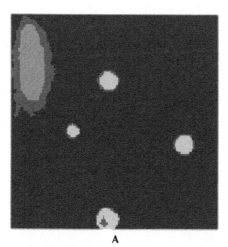

A

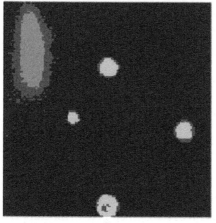

B

FIGURE 2.16

Results obtained with the SVD preprocessing and BP classifiers. (A) The results on the image reconstructed with the components S_1. (B) The results on the image reconstructed with the components S_1 and S_2.

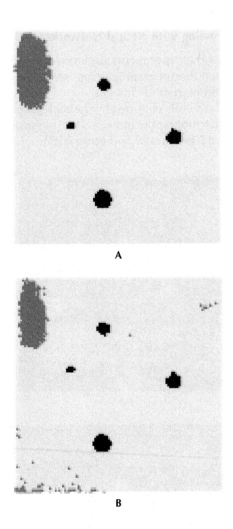

FIGURE 2.17
Results obtained with the SVD preprocessing and SOM classifiers. (A) The results on the image reconstructed with the component S_1. (B) The results on the image reconstructed with the components S_1 and S_2.

Figure 2.17 shows the results obtained by the SOM classifier applied to the thermographic images after the reconstruction with the components S_1 (Figure 2.17A) and S_1S_2 (Figure 2.17B). Experiments demonstrated that the first components of the SVD decomposition actually contained the largest information content of the signal, whereas the image reconstruction by using the successive components did not increase the quality of the signal but introduced noise in the final segmentation. To confirm this assessment, we plotted (Figure 2.18) the results obtained using a SOM classifier after the reconstruction with $S_1S_2S_3$ (Figure 2.18A) and

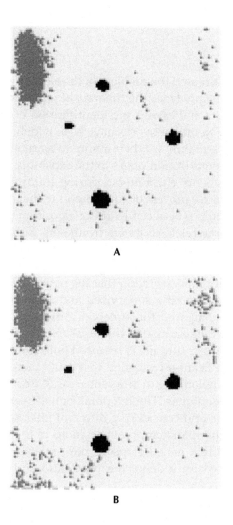

FIGURE 2.18
Results obtained with the SVD preprocessing and SOM classifiers. (A) The results on the image reconstructed with the components S_1, S_2, and S_3. (B) The results on the image reconstructed with the components S_1, S_2, S_3, S_4, S_5, S_6, and S_7.

with $S_1S_2S_3S_4S_5S_6S_7$ (Figure 2.18B). Comparing the results in Figures 2.18 and 2.17, it was evident that by using just the first component S_1 to reconstruct the image sequence, the noise was greatly reduced and defect areas were more clearly visible than in the following experiments with more components. Also, in these experiments, the BP classifiers produced a more effective segmentation image, but they required human intervention for the manual selection of training samples. SOM classifiers were less precise in space localization of defective areas but remained more attractive for the few a priori knowledge required.

2.5 Conclusions

In this chapter we addressed the problem of developing an automatic signal processing system that analyzes the time/space variations in a sequence of thermographic images and allows the identification of internal defects in composite materials. Neural networks have been widely applied for pattern recognition problems, mainly for their ability to learn and therefore generalize knowledge. Generalization refers to the capability to produce reasonable outputs for inputs not encountered during training. Neural networks learn through an iterative process of adjustments to the internal weights and thresholds. The difference between learning algorithms lies in the manner in which the networks relate to its environments. Supervised paradigms require the a priori knowledge of a set of input–output examples; unsupervised learning paradigms learn to organize the input data using only a measure of the quality representation that the network is required to learn. In this chapter we compared a supervised and an unsupervised learning approach. Also, we compared the effects of different preprocessing techniques by evaluating the results after the supervised and unsupervised classifiers. A preprocessing technique is required both for noise reduction and for extraction of significant information that could increase the probability of separating signals belonging to different areas. We considered two different preprocessing techniques. The temporal signals were filtered applying the wavelet transform and considering different filter bank configurations. Then, we analyzed the spatiotemporal variations of the thermographic signals as we observed that neighbor pixels have related temperature variations that can be used for a denoising procedure on the original signals (SVD preprocessing).

Experiments demonstrated that the SVD preprocessing gave lightly better results than the wavelet filtering. These results were expected because the SVD allowed signal reconstruction considering the space and time variation of neighboring points, not only the temporal variation as with the wavelet preprocessing. Comparing the results of the two classifiers, we observed a higher quality of the image segmentation with the BP classifiers. However, the results obtained with the SOM classifiers are very promising because they do not require any manual selection of training samples and are able to aggregate data considering only the signal similarity. In our opinion, neural approaches for defect recognition can produce great advantages, increasing the speed of extensive and exhaustive checks and also providing supports to human operators that have to concentrate their attention only when anomalies are detected by automatic systems. The proposed methodologies could be successfully applied also to other signals for nondestructive inspection.

References

Avdelidis, N. P., Hawtin, B. C., and Almond, D. P. 2003. Transient thermography in the assessment of defects of aircraft composites. *NDT&E International* 36:433–439.

D'Orazio, T., Guaragnella, C., Leo, M., and Spagnolo, P. 2005. Defect detection in aircraft composites by using a neural approach in the analysis of thermographic images. *NDT&E International* 38:664–673.

Freeman, J., and Skapura, D. 1991. *Neural network algorithms, applications, and programming techniques.* Boston, MA: Addison-Wesley.

Galietti, U., Luprano, V., Nenna, S., Spagnolo, L., and Tundo, A. 2007. Non-destructive defect characterization of concrete structures reinforced by means of FRP. *Infrared Physics & Technology* 49:218–223.

Gaussorgues, G. 1994. *Infrared thermography.* London: Chapman & Hall.

Giorleo, G., Meola, C., and Squillace, A. 2000. Analysis of detective carbon-epoxy by means of lock-in thermography. *Research in Nondestructive Evaluation* 12:241–250.

Haykin, S. 1994. *Neural network: a comprehensive foundation.* New York: Macmillan and IEEE Press.

Huang, Y. D., Froyen, L., and Wevers, M. 2001. Quality control and nondestructive test in metal matrix composites. *Journal of Nondestructive Evaluation* 20:113–132.

Inagaki, T., Ishii, T., and Iwamoto, T. 1999. On the NDT and E for the diagnosis of defects using infrared thermography. *NDT&E International* 32:247–257.

Jones, T. S. 1995. Infrared thermographic evaluation of marine composite structures. *SPIE* Vol. 2459:42.

Just-Agosto, F., Serrano, D., Shafiq, B., and Cecchini, A. 2008. Neural network based nondestructive evaluation of sandwich composites, *Composites Part B: Engineering* 39:217–225.

Maldague, X., Largouët, Y., and Couturier, J. P. 1998. A study of defect using neural networks in pulsed phase thermography: modeling, noise, experiments. *Revue Générale de Thermique* 37:704–717.

Marin, J. Y., and Tretout, H. 1999. Advanced technology and processing tools for corrosion detection by infrared thermography. *AITA–Advanced Infrared Technology and Applications* 128–133.

Saintey, M. B., and Almond, D. P. 1997. An artificial neural network interpreter for transient thermography image data. *NDT&E International* 30:291–295.

Sakagami, T., and Kubo, S. 2002. Applications of pulse heating thermography and lock-in thermography to quantitative nondestructive evaluations. *Infrared Physics & Technology* 43:211–218.

Vallerand, S., and Maldague, X. 2000. Defect characterization in pulsed thermography: a statistical method compared with Kohonen and Perceptron neural networks. *NDT&E International* 33:307–315.

Wu, D., and Busse, G. 1998. Lock-in thermography for nondestructive evaluation of materials. *Revue Generale De Thermique* 37:693–703.

3

The Use of Artificial Neural Networks in Damage Detection and Assessment in Polymeric Composite Structures

S. John, A. Kesavan, and I. Herszberg

CONTENTS

ABSTRACT The complexity of damage detection, assessment, and prognosis has, in recent years, led to the search for data evaluation tools such as artificial intelligence tools to enable the identification of trends and features of a structural response to damage. This sort of feature in any structural health monitoring (SHM) system has always been desirable because of the robustness characteristics required in any deployable SHM. The work presented in this chapter outlines in detail the research methodology used to investigate the use of artificial neural networks (ANNs) applied to static strain response on a physical structural member. A suite of preprocessing algorithms was developed for the ANN to function adequately. A brief description of the threefold validation technique used to decide on the type of learning rule and network architecture is given. The entire study was conducted on a thick-sectioned glass-reinforced polymer composite T-joint—one similar to that used in full-scale marine hull construction. The finite element analysis was also conducted by placing delaminations of different sizes at various locations in two structures, a composite beam and a T-joint. Glass fiber-reinforced polymer T-joints were then manufactured and tested, thereby verifying the accuracy of the finite element analysis results experimentally. The SHM system was found to be capable of not only detecting the presence of multiple delaminations in a composite structure but also determining the location and extent of all the delaminations present in the T-joint structure, regardless of the load (angle and magnitude) acting on the structure. This SHM system necessitated the development of a novel preprocessing algorithm, Damage Relativity Assessment Technique, along with a pattern recognition tool, ANN, to predict and estimate the damage. Another program developed—the Global Neural Network Algorithm for Sequential Processing of Internal Sub Networks—uses multiple ANNs to render the SHM system independent to variations in structural loading and capable of estimating multiple delaminations in composite T-joint structures. Up to 82% improvement in detection accuracy was observed when the Global Neural Network Algorithm for Sequential Processing of Internal Sub Networks was invoked. The resulting strain distribution from the finite element analysis was preprocessed by the Damage Relativity Assessment Technique and used to train the ANN to predict and estimate damage in the structures. Finally, on testing the SHM system developed with strain signatures of composite T-joint structures, subjected to variable loading, embedded with all possible damage configurations (including multiple-damage scenarios), an overall damage (location and extent) prediction accuracy of 94.1% was achieved. These results are presented and discussed in detail in this publication.

3.1 Need for Artificial Neural Networks in Structural Health Monitoring

This field is of paramount importance especially with the increase in use of composites in structures. The increase in use of composites is obvious because of its distinct advantages, namely, high strength-to-weight ratio, high stiffness, high corrosion resistance, etc. Unfortunately, these composites are prone to in-service defects like matrix cracking, delamination, fiber breakage, and debonding. This necessitates the need for an online SHM system, which is capable of determining the presence of the damage (in the incipient stage itself), the location of the damage, the size of the damage, and possibly the criticality of the damage. Various nondestructive techniques, which are capable of achieving the goal, have been discussed by several researchers such as Yan et al. (2007) and Kesavan et al. (2008). For complex situations (variable loading, variable damage level, and variable type of damage) using complex structures, the response signal obtained from the sensors (damage signature vector) will be complicated. This makes it difficult to decode the signal to determine the damage location and the damage. Moreover, it is also difficult or impossible to create accurate mathematical models for complex structures due to both geometric and material property nonlinearity. This is where the role of data evaluation techniques such artificial neural networks (ANNs) might be pivotal in identifying phenomenological trends or patterns. The following features of neural networks make it an effective tool in SHM.

Generalization ability: Generalization (Haykin, 1999) refers to the neural network producing reasonable outputs for inputs not encountered during training (learning). This information-processing capability makes it possible for neural networks to solve complex (large-scale) problems that are currently intractable. Hence, for ANNs to detect damage, the network has to be trained with sufficient damage signatures and their corresponding physical parameters.

Nonlinearity: Artificial neural networks can be used even for nonlinear problems (Hagan et al., 1995; Haykin, 1999; The Mathworks, 1999), as the interconnected neurons can be either linear or nonlinear. Multiple interconnections between neurons tend to make the system nonlinear. This feature is extremely important in the field of SHM as the signals from complex structures under variable loading may be nonlinear.

Input-output mapping: This is the most powerful feature of the neural network that involves supervised learning. The network tries to correlate a unique input signal with a desired response. It modifies the synaptic weights by a learning process to achieve the desired response. Training the network involves feeding the network with a set of input signals and the corresponding desired response. The network then tries to learn from the examples by constructing an input-output mapping (Hagan et al., 1995; Haykin, 1999; The Mathworks, 1999) for the problem at hand.

Adaptivity: Neural networks adapt easily to changes in the environment, by adjusting the synaptic weight accordingly on retraining the system. Real-time networks that are capable of changing its synaptic weights automatically can be designed for nonstationary environmental conditions. This feature clubbed with the natural network architecture is quintessential for adaptive pattern classification, adaptive signal processing, and adaptive control (Haykin, 1999).

Evidential response: For pattern classification purposes, networks can be designed to provide information about the pattern to be selected, as well as the confidence of the decision made. This helps in rejecting indistinct patterns, thereby improving the classification performance of the network.

Contextual information: Related information is dealt with naturally by the network as the knowledge is represented by the very structure and activation state of the network. Essentially, every neuron is affected by the activity of all other neurons in the network.

Fault tolerance: Neural networks are inherently fault tolerant. If the neural networks were implemented in hardware form, and if a neuron or a connecting link is damaged, then only the output quality deteriorates (Haykin, 1999) rather than the system failing completely. This is mainly because of the distributed nature of the information stored in the network; the damage to the hardware has to be extensive for the network to fail completely.

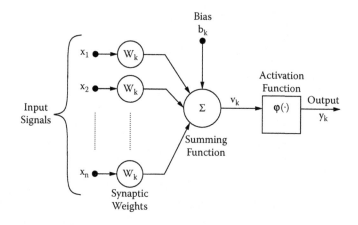

FIGURE 3.1
Nonlinear model of a neuron.

Scores of researchers have documented the use of ANNs in tandem with existing nondestructive testing techniques for the purpose of SHM. Yam et al. (2003) effectively made use of ANNs to establish a mapping relationship between the structural vibration response and the structural damage status. Seo and Lee (1999) studied the relationship between the electrical resistance change, fatigue life, and stiffness reduction in carbon fiber-reinforced plastics making use of neural networks. Su and Ye (2005) have documented the use of neural networks to localize the damage from lamb wave response signals. Neural networks were trained with global image data, by Nyongesa et al. (2001), obtained from stereographic imaging of laminated composites. Nyongesa et al. (2001) made use of this trained network to detect impact damage on composites. Similar work has also been published by Orazio et al. (2005). In this, the neural network was trained with thermographic images of composites with subsurface cracks and was then used for damage detection purposes.

3.1.1 Neuron Model

A neuron is the information-processing unit of the neural network, much like the brain in human beings (Hagan et al., 1995; Haykin, 1999; The Mathworks, 1999). Figure 3.1 shows the block diagram of the neuron.

The neuron consists of three main parts: a set of *synapses*, which connect the input signal (x_j) to the neuron via a set of weights (W_{kj}); an *adder* (u_k), which sums up the input signals, weighted by the respective synapses of the neuron; and an *activation function* $[\varphi(\circ)]$ for limiting the amplitude of the output of the neuron. At times, a bias (b_k) is added to the neuron to increase or decrease the net output of the neuron.

Mathematically (Haykin, 1999), a neuron k is described as

$$u_k = \sum_{j=1}^{n} w_{kj} x_j \tag{3.1}$$

$$y_k = \phi(u_k + b_k) \tag{3.2}$$

where

$x_1, x_2, x_3, \ldots, x_n$ are the input signals,
$w_{k1}, w_{k2}, \ldots, w_{kn}$ are the weights for neuron k,
b_k is the bias,
u_k is the adder or the linear combiner,
$\varphi(\circ)$ is the activation function,
y_k is the output signal of the neuron.

The output range of the neuron depends on the type of activation function used. There are four types activation functions (The Mathworks, 1999) commonly used, namely, hard-limit activation function, log-sigmoid activation function, tan-sigmoid activation function, and linear activation function.

3.1.2 Activation Function

3.1.2.1 *Hard-Limit Activation Function*

The hard-limit activation function is shown in Figure 3.2. This function limits the output (y_k) of the neuron to either 0, if the net input argument v_k is less than zero, or 1, if v_k is greater than or equal to zero.

Mathematically, the hard-limit activation function can be described as:

$$y_k = \begin{cases} 1 & \text{if } v_k \geq 0 \\ 0 & \text{if } v_k < 0 \end{cases} \tag{3.3}$$

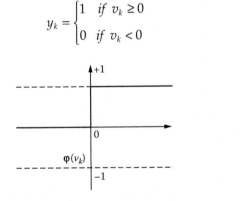

FIGURE 3.2
Hard-limit activation function.

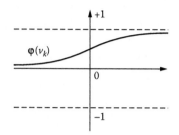

FIGURE 3.3
Log-sigmoid activation function.

3.1.2.2 Log-Sigmoid Activation Function

The sigmoid activation function is shown in Figure 3.3. This function limits the output (y_k) to a range of 0 to 1, according to the expression:

$$y_k = \frac{1}{(1+e^{-v_k})} \tag{3.4}$$

The input can be anything between plus and minus infinity.

3.1.2.3 Tan-Sigmoid Activation Function

The tan-sigmoid activation function is shown in Figure 3.4. This function takes the input, which again may have any value between plus and minus infinity, and limits the output into the range −1 to 1.

This is done according to the following mathematical expression:

$$y_k = \tanh(v_k) \tag{3.5}$$

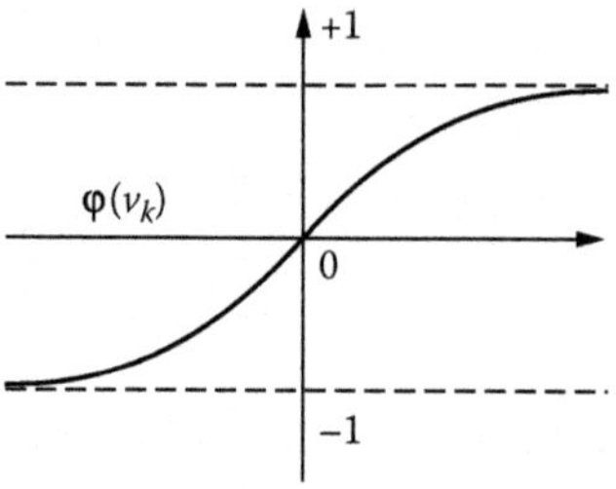

FIGURE 3.4
Tan-sigmoid activation function.

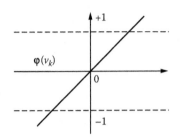

FIGURE 3.5
Linear activation function.

3.1.2.4 Linear Activation Function

The linear activation function shown in Figure 3.5 reproduces the input fed into it as the output itself.

3.1.3 Network Architecture

3.1.3.1 Single-Layer Feed-Forward Network

The simplest form of the feed-forward network is a single-layer network (Haykin, 1999) shown in Figure 3.6. The network consists of the input layer consisting of source nodes directly connected to the output layer, with neurons (computational nodes). This type of a network has no hidden layers.

3.1.3.2 Multilayer Feed-Forward Network

The feed-forward network consisting of one or more hidden layers between the input and the output layer is called a multilayer feed-forward network

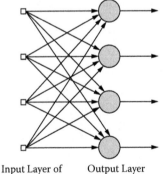

Input Layer of   Output Layer
Source Nodes   of Neurons

FIGURE 3.6
Single-layer feed-forward network.

(Haykin, 1999). A typical multilayer network is shown in Figure 3.7, consisting of an input layer with 10 nodes, one hidden layer with four neurons, and an output layer with two neurons. The presence of the hidden layer neurons enables the network to extract higher-order statistics, which tends to be useful when the input layer is large.

A number of researchers have documented the use of multilayer feed-forward networks. Altinkok and Koker (2006) made use of the multilayer feed-forward network to determine the properties of metal matrix composites, whereas Jia and Davalos (2006) made use of the same for predicting the fatigue crack growth in fiber reinforced plastics bonded with wood interfaces. Multilayer networks were also used by Ishak et al. (2001) for localizing delaminations in composite laminate beams using its displacement responses.

3.1.4 Types of Networks and Training Algorithms

3.1.4.1 Feed-Forward Back-Propagation Network

Feed-forward back-propagation network (Dayhoff, 1990; Hagan et al., 1995; Haykin, 1999; The Mathworks, 1999; Simpson, 1990) basically consists of three layers: the input, output, and one or more hidden layers. Each layer consists of multiple neurons. Sigmoid or tan-sigmoid activation function is normally used in the hidden layers, and linear activation functions are used in the output layers. These networks are also known as multilayer perceptrons.

These networks are trained using algorithms known as error back-propagation algorithms (Haykin, 1999). Error back-propagation learning can be divided into two stages, namely, the forward pass and the backward pass.

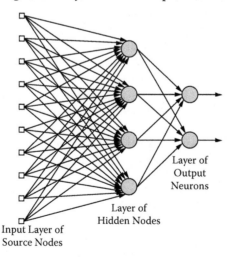

FIGURE 3.7
Multilayer feed-forward network.

During the forward pass, the input vector (damage signature) is applied to the input nodes of the network. This signal propagates through the network layer by layer and finally produces an actual output. In this forward pass stage, the synaptic weights are not modified and they remain constant. Once actual output is obtained, it is compared with the target output (damage size) and an error signal is obtained. This error signal is then propagated back into the network in the opposite direction. During this stage, the synaptic weights are adjusted in such a way that the actual response moves closer to the target response. Properly trained back-propagation networks tend to give reasonable answers when presented with inputs that they have never seen. Typically, a new input leads to an output similar to the correct output for input vectors, which are similar to the ones used for training. This generalization property makes it possible to train a network on a representative set of input/target pairs and get good results without training the network on all possible input/output pairs. A multilayer feed-forward back-propagation network is shown in Figure 3.8. This network consists of an input layer, two hidden layers, and an output layer,
where

X_n is the input vector,

$w_{k,n}$ is the weight matrix,

$$v_k = \sum_{i=0}^{n} (w_{k,n} y_i + b_k)$$

$\varphi(-)$ is the activation function,

$y_k = \varphi(v_k)$ is the output.

A lot of work has been reported in the area of SHM using the feed-forward back-propagation network. Xu et al. (2001) made use of the feed-forward network for damage detection in anisotropic laminated plates. Feed-forward networks were used by Graham et al. (2004) to detect impact-induced damage in tandem with an eddy current-based technique. Ativitavas et al. (2006) successfully predicted the different types of failure mechanisms in fiber-reinforced composites using acoustic emission response as input to the feed-forward back-propagation network. These networks were also used by Lee and Yun (2006) for detecting damage on bridges using the ambient vibration data as its input.

3.1.5 Training Algorithm

The main objective of all the training algorithms is to adjust the synaptic weights in such a way that the performance function decreases. The performance function is a quantitative measurement of network performance.

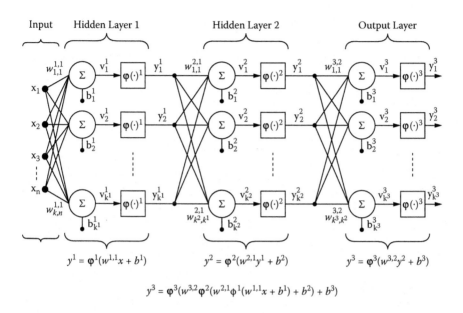

FIGURE 3.8
Multilayer feed-forward back-propagation network.

Therefore, if the network performs well, the performance function is small; whereas if the network performs poorly, the performance function is large. A sample of three such algorithms is discussed in the following subsections.

3.1.5.1 Gradient Descent Algorithm

The gradient descent algorithm (Hagan et al., 1995; The Mathworks, 1999) updates the network weights and biases in the direction in which the performance function decreases most rapidly (negative of the gradient). An iteration of this algorithm is shown below:

$$x_{k+1} = x_k - \alpha_k g_k \qquad (3.6)$$

where the current weight vector is represented by x_k, the current gradient is g_k, and α_k is the learning rate. The current weight vector is added to the product of the learning rate and the negative of the gradient to determine the new weight. The learning rate determines the size of the step. If the learning rate chosen is too large, then the algorithm becomes unstable and does not converge; whereas if the learning rate chosen is too small, then the algorithm takes too long to converge.

Incremental mode and batch mode are the two ways in which the gradient descent algorithm can be implemented. In the incremental mode, the gradient is computed and the weights are updated after each input is applied to the network. In the batch mode, all of the inputs are applied to the network before the weights are updated. Su and Ye (2004) and Fang et al. (2005) have documented the use of the steepest descent algorithm for training the feed-forward network to detect damage in a composite laminate and a cantilevered beam, respectively.

3.1.5.2   Gradient Descent with Momentum

This algorithm makes use of a modified version of the gradient descent algorithm. This algorithm adds a momentum factor to the gradient decent algorithm, thereby providing faster convergence. The momentum makes the network ignore all the small features in the error surface. Without momentum, the network may get stuck in shallow local minimums (Hagan et al., 1995; The Mathworks, 1999).

Momentum is added to the gradient descent algorithm by making weight changes equal to the sum of a fraction of the last weight change and the new weight change, where the new weight is calculated by the gradient descent algorithm. The momentum constant, which has a value between 0 and 1, determines the effect that the last weight change is allowed to have on the current weight update. If the momentum constant is 0, then the last weight change is not taken into account at all. However, if the momentum constant is 1, then the new weight change is equated to the last weight change and the gradient is not taken into account.

3.1.5.3   Resilient Back-Propagation Algorithm

The gradient decent algorithms are not suitable for using with multilayer networks. This is mainly because of the fact that these networks typically use the sigmoid or the tan-sigmoid activation functions. The slope of the sigmoid and the tan-sigmoid functions approach zero when the input is large. This effect causes the weights to be updated by only a minute fraction, when using the gradient descent algorithm, although they are far from their optimal value.

The resilient back-propagation algorithm (Hagan et al., 1995; The Mathworks, 1999) considers only the sign of the gradient to determine the direction of the weight update. The size of the weight change is governed by a separate update value. This method therefore eliminates the effects of the magnitudes of partial derivatives.

In this method, if the derivative of the performance function has the same sign for two successive iterations, with respect to the current weight, then the update value is increased. The update value is decreased in case the derivative with respect to the current weight change signs from the previous iteration. The update value remains constant if the derivative is zero. If the weights tend to oscillate, then the weight change will be reduced. And, finally, if the weight continues to change in the same direction for several iterations, the magnitude of weight change is increased.

This algorithm is more robust and facilitates faster convergence than the steepest descent algorithms; hence, it is apt for multilayer feed-forward networks. However, not many researchers have made use of this algorithm for training ANNs. Resilient back propagation was used by Kesavan et al. (2005) to determine the location and size of delaminations in composite ship structures.

3.2 Damage Relativity Assessment Technique

The Damage Relativity Assessment Technique (DRAT) is a preprocessing program developed that was initially coded in Java, the object-oriented programming language. The main function of the DRAT is to identify the values pertaining to the delamination and filter out the remaining. The DRAT was initially written for the case of a beam experiencing three-point beam, with a constant magnitude of load.

In the composite beam case, the DRAT protocol compares the surface strain distribution of the healthy beam with that of the damaged beam. It then computes the difference of these two vectors and feeds the output to a filtered damage signature database (DSD). This process is then carried out for all the surface strain signature vectors present in the DSD. The filtered DSD is then used to train the ANN. The DRAT filtration process is shown in Figure 3.9. The filtered value obtained is similar to the strain anomaly observed along the bond line, in the presence of damage. This preprocessing algorithm was then used to filter the DSD used for training the network. The network with 23 sensory nodes in the input layer, 12 neurons in the hidden layer, and 1 neuron in the output layer, which was used to predict the damage location with the bond-line strain distribution, was used in this case, too. The performance function vs. iteration and the training error vs. number of training cases obtained on training the network with the filtered DSD are shown in Figures 3.9 and 3.10, respectively.

Figure 3.11 shows the comparison of training errors obtained by the network before and after the DRAT filtration process. Figure 3.11 clearly demonstrates the extent to which the DRAT reduces the training errors. It also highlights the importance of the DRAT in preprocessing the strain signatures.

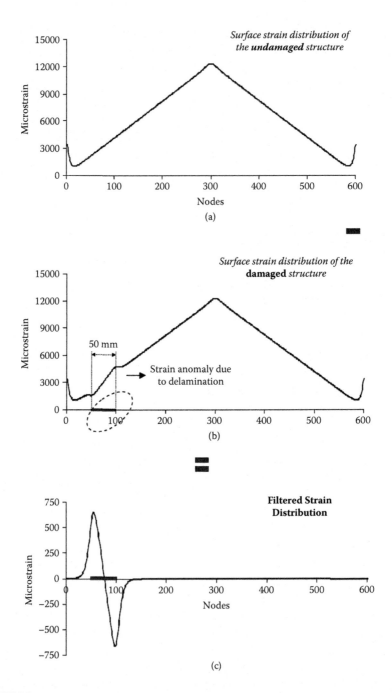

FIGURE 3.9
The DRAT process is illustrated, where (a) is the bottom surface strain distribution of an undamaged beam, (b) is the bottom surface strain distribution of a beam embedded with a 50-mm delamination, and (c) is the filtered value obtained using the DRAT.

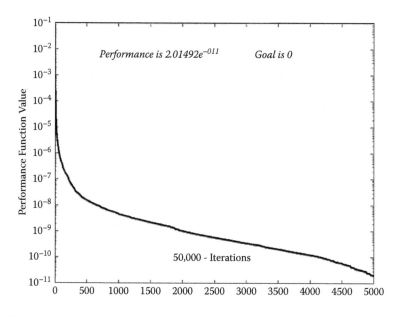

FIGURE 3.10
Training curve obtained on using the filtered strain signatures, obtained from the surface of the beam, to train the 12-neuron network.

3.3 Introduction to Detection of Damage in the Horizontal Section of a Composite T-Joint

The ability of the neural network to detect and determine the extent of damage in complex structures, such as the composite T-joint structure, was initially tested with the horizontal damage scenario. The composite T-joint used in this study was a glass-fiber Vinyl ester composite. Kesavan (2006) and Kesavan et al. (2008) describe the manufacture of the T-joints and the experimental tests conducted, respectively. In this damage configuration, delaminations were located between the overlaminates and the hull section and also between the filler and the hull section as shown in Figure 3.12. The size of the delaminations embedded in the model, for training, ranged from 30 to 100 mm. These delaminations were placed anywhere between the two constraints. The T-joint was then subjected to a tensile load of 5 kN; the strain distribution obtained from along the bond line and along the surface (overlaminates) was then tabulated.

3.3.1 Sensors along the Horizontal Bond Line

Eleven sensors were placed along the horizontal bond line as shown in Figure 3.12. The spacing between the sensors varied from 50 to 10 mm. The

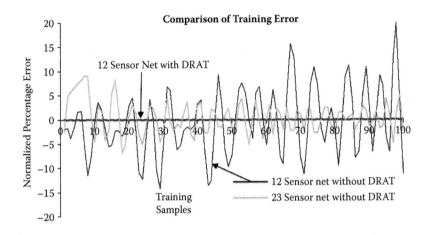

FIGURE 3.11
Comparison of the training errors obtained by the network before and after the DRAT filtration process.

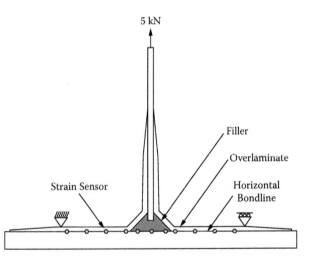

FIGURE 3.12
Bond-line sensor configuration.

spacing between the sensors, near the center of the hull, is decreased to avoid errors when the delaminations are located beneath the loading point, similar to the beam model case.

A PATRAN command language program was then written to hasten the T-joint and delamination modeling process. The strain output obtained from 11 nodes corresponding to the sensors was then obtained for each damage case and inserted into the DSD. This DSD was then used to train the ANN.

A separate test set was also modeled, which consisted of strain signatures of T-joints embedded with delaminations of sizes and positions that were not a part of the training set.

3.3.2 Network Architecture

The architecture of the network used for predicting the presence and estimating horizontal delaminations in T-joints is shown in Figure 3.13. This network consisted of an input layer with 11 sensory nodes (corresponding to the location of the strain sensors); three hidden layers with 11, 25, and 15 neurons, respectively; and an output layer with 2 neurons (to predict the location and extent of damage).

The activation function used in the hidden layers was tan-sigmoid; and for the output layer, log-sigmoid was used. In this case the resilient back-propagation algorithm was used to train the network, as this network was unable to converge using the scaled conjugate gradient training algorithm.

The number of iterations was set to a maximum of 75,000 and the minimum gradient for early stoppage was set as $10e^{-10}$. An optimization method called the threefold cross-validation test was used to select the number of neurons in the hidden layers.

3.3.3 Threefold Cross-Validation Test

The generalization power of the ANN reduces drastically if the appropriate number of neurons for every layer is not selected. To eliminate this problem, the network is optimized using a threefold cross-validation technique. This technique involves the division of the entire training set into three equal parts (A, B, C). The number of neurons in the appropriate network is initially selected at random. This network is then trained using A and B as

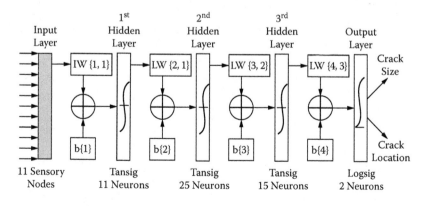

FIGURE 3.13
Network architecture for detecting horizontal delamination embedded in T-joints.

the training set and is tested using C; the output is recorded. The process is again repeated using the combinations B and C for training and A for testing and finally C and A for training and B for testing.

If the training error produced by the three different cases is above the acceptable limit, the number of neurons in the hidden layers is increased and the entire process is repeated. This process is described in detail in Kesavan (2006).

3.4   Introduction to the Modified DRAT: A Preprocessor to the Global Neural Network Architecture for Incorporating Sequential Processing of Internal Sub Networks

Good agreement between the experimental and the computational results facilitated the use of finite element modeling for damage detection purposes. The intent of this chapter was to see if the ANNs combined with the preprocessing algorithm DRAT was capable of detecting delaminations (single or multiple) embedded in T-joints, regardless of the load applied on the T-joint.

The FE models created were embedded with delaminations of different sizes at different locations. Sensors were then placed on the surface of the T-joint to sense the strain, which was then processed and inputted into the ANNs for detecting the presence and estimating the extent of the damage. To detect multiple delamination, a Global Neural Network Architecture Incorporating Sequential Processing of Internal Sub Networks (GNAISPIN) was developed, which was capable of virtually combining multiple ANNs into a single global network.

3.4.1   Damage Configurations Considered

A stress analysis of the T-joint model was conducted to identify the critical areas, prone to damage. Figure 3.14 shows the stress distribution (X-component) across a healthy T-joint. From the figure, it is clear that stress concentration occurs around the edges (a, b, c, d) of the filler section. Hence, it is assumed that when a delamination occurs, it is likely to initiate from one of these corners.

A total of eight critical damage positions were identified and modeled:

- Delamination between the overlaminate and the hull, to the left or right side of the bulkhead (LH or RH delamination).
- Delamination between the overlaminate and the filler section, to the left or the right side of the bulkhead, initiating from the bottom corner of the filler (SLB or SRB delamination).

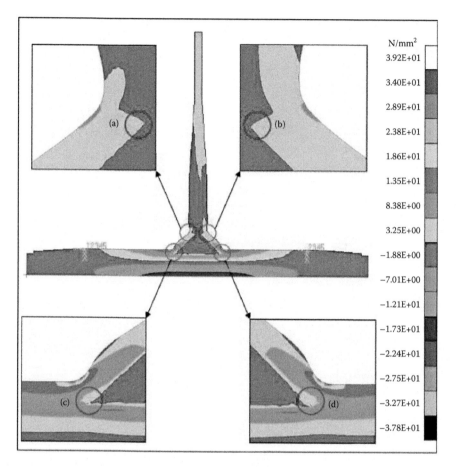

FIGURE 3.14
Stress distribution on a healthy T-joint in tension.

- Delamination between the overlaminate and the filler section, to the left or the right side of the bulkhead, initiating from the top corner of the filler (SLT or SRT delamination).
- Delamination between the overlaminate and the bulkhead, on the left or right side of the bulkhead (LV or RV delamination).

Figure 3.15 shows the critical damage configurations considered and modeled for this part of the study.

3.4.2 Sensor Configuration

It was determined that eight sensors (on the surface of the structure) were sufficient to predict the presence of damage, identify its location, and also estimate the extent of damage in all the damage configurations discussed. The locations

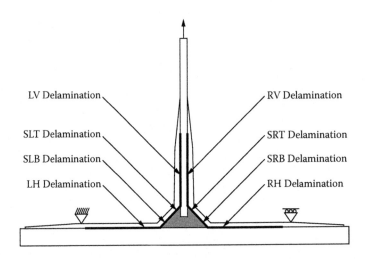

LV Delamination

RV Delamination

SLT Delamination

SRT Delamination

SLB Delamination

SRB Delamination

LH Delamination

RH Delamination

FIGURE 3.15
Various critical damage configurations.

of these sensors are shown in Figure 3.16. The strain obtained from these sensors is then preprocessed and used as input to the ANN. A couple of reference sensors were required to remove the effect of load acting on the structure, as shown in Figure 3.16. These reference sensors were placed at a location where there was minimal probability of the occurrence of a delamination.

3.4.3 Network Architecture

The architecture of the network used to detect the presence, the location, and the extent of damage, for the various damage configurations, in a T-joint is similar to that in Figure 3.13. This network consisted of an input layer with eight sensory nodes (corresponding to the location of the strain sensors); three hidden layers with eight, seven, and five neurons, respectively; and an output layer with eight neurons (to predict the location and extent of damage). Tan-sigmoid activation function was used in the first and the second hidden layers and log-sigmoid was used in the third hidden layer and the output layer. In this case, the resilient back-propagation algorithm was used to train the network, as this network was unable to converge when using the scaled conjugate gradient training algorithm. The number of iterations was set to a maximum of 75,000 and the minimum gradient for early stoppage was set as $1e^{-10}$. The threefold cross-validation test was used to select the number of neurons in the hidden layers.

The filtered damage signature test database (DSTD) was then keyed into the trained neural network. The simulated response obtained from the neural network is presented in Table 3.1. Table 3.1 shows a comparison of the actual and the predicted delamination size. The normalized prediction error is represented as:

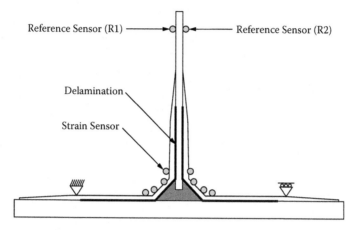

Reference Sensor (R1) ——————⊸ ⊶—————— Reference Sensor (R2)

Delamination

Strain Sensor

FIGURE 3.16
Location of the sensors on the surface of the T-joint.

$$\text{normalized percentage error} = \left(\frac{\text{actual size} - \text{predicted size}}{\text{max. delamination size}} \right) 100$$

where the maximum length of the delamination is taken as 100 mm.

From Table 3.1, it is evident that the average prediction accuracy of the system was found to be 99.3%. The SHM system was found to be capable of predicting the size of the delamination in all the cases studied, regardless of the loading magnitude.

The network also predicts the location of the delamination (LH, RH, LV, RV, SLB, SLT, SRB, or SRT) precisely in all the cases; hence, the percentage error is not tabulated.

It is evident from the above discussion that the SHM system developed was capable of predicting and estimating the damage regardless of the magnitude of the load acting on the system. Therefore, the logical next step would be to test its performance when the angle at which the load applied is varied.

3.4.4 Variable Loading Angle and Magnitude

3.4.4.1 *Training Set*

To make the SHM system insensitive to the loading angle and the magnitude of load acting on the structure, a few changes had to be made to the training set and also to the DRAT protocol. As an initial assumption, the network and sensor architecture used in the previous case study (SHM insensitive to loading magnitude) was used again. The network consisted of four layers; the input layer had eight sensory nodes and three hidden layers with eight, seven, and five neurons, respectively; and the output layer had eight neurons (similar to Figure 3.13).

TABLE 3.1

Actual and Predicted Damage Size and Location in a T-Joint Structure Subjected to Tensile Loading of Different Magnitudes

Load	Delamination Location (Labels Described in Figure 10.2)	Size of the Delamination			
		Actual	Predicted	Difference	Normalized Error
N		mm	mm	mm	%
6000	LH	15	14.8	0.2	0.2
2500	LH	19	18.8	0.2	0.2
3000	RH	24	23.8	0.2	0.2
3750	RH	29	29.6	0.6	0.6
5000	SLB	33	33.5	0.5	0.5
4500	SLB	37	37.9	0.9	0.9
2750	SLT	39	41.4	2.4	2.4
7000	SLT	42	43.9	1.9	1.9
6500	SRT	46	44.8	1.2	1.2
4750	SRT	49	49.1	0.1	0.1
2500	SRB	51	52.1	1.1	1.1
3250	SRB	53	51.4	1.6	1.6
5750	LV	64	63.4	0.6	0.6
5250	LV	69	68.8	0.2	0.2
8000	LV	76	76.5	0.5	0.5
8500	RV	85	85.4	0.4	0.4
9000	RV	94	94.5	0.5	0.5
8750	RV	97	97	0	0
−8500	RV	85	85.4	0.4	0.4
−5000	RV	85	85.4	0.4	0.4
			Average error		0.7%

The training set used in the previous case study, consisting of filtered damaged strain signatures of T-joints embedded with delamination subject to a pull-off load of 5 kN at a loading angle of 0.55°, was normalized. The maximum absolute strain value for every damage strain signature, in the training set, was identified; and it was used to normalize the corresponding strain signature. This normalized training set was then used to train the ANN.

3.4.4.2 Test Set

A test set was then created, which consisted of T-joints embedded with delaminations of sizes that were again not a part of the training set. The magnitude of the load applied and the loading angle were also varied for every test model. The strain signatures obtained from these models were entered into the DSTD. The DRAT algorithm had to be further modified to remove the effect of the loading magnitude as well as loading angle.

3.4.5   DRAT Modification

The DRAT algorithm had to be modified to remove the effect of the magnitude and the loading angle. First, a database of strain signatures (undamaged) obtained from healthy T-joints loaded at a constant magnitude of 5 kN, and variable loading angles were created. These healthy T-joints also consisted of the reference sensors, R3 and R4, located corresponding to sensors R1 and R2.

Next, the ratio R3/{(Abs(R3) + Abs(R4)} is calculated for all the cases in the healthy database. A polynomial curve fitting was then performed for strains obtained from every sensor location for different loading angles. The polynomial equation is represented as:

$$y = a_0 + a_1 x + ... + a_k x^k \tag{3.7}$$

where

$$x = \frac{R3}{Abs(R3) + Abs(R4)} \tag{3.8}$$

$y = \varepsilon_i$ (where i indicates the sensor no.)
a_k = coefficient of the polynomial
k = order of the polynomial

A total of 10 polynomial equations were created, that is, for every sensor location. The polyfit function in Matlab® was then used to obtain the coefficients of the polynomial equations created. These polynomial equations were then used to determine the healthy strain signature of a T-joint experiencing an identical loading pattern as the test case (T-joint embedded with damage). The ratio R1/{(Abs(R1) + Abs(R2)} of the damaged strain signature from the test case under investigation was then computed. This ratio was then substituted into all the 10 polynomial equations created to determine the strain signature of a healthy T-joint experiencing an identical loading pattern as the test case. The ratio of the strain, obtained at the reference sensors R3 and R1, corresponding to the healthy structure and the damaged structure was then computed. The result obtained was factored with the damaged strain signature. The damaged strain signature was then compared with the estimated healthy strain signature, and the unnecessary features were removed to obtain a filtered strain signature.

The maximum absolute strain value of the filtered strain signature was identified and it was used to normalize the strain signature. The normalized strain signature was then inserted into the DSTD. This DSTD was then used to test the performance of the ANN.

3.4.5.1   Performance of the Modified DRAT

A schematic of the modification made to the DRAT is shown in Figure 3.17. Figures 3.18, 3.19, and 3.20 show the effect of using the modified DRAT (MDRAT) algorithm to filter the damage signatures. Figure 3.18 shows the actual damage signature, the DRAT filtered damage signature, and the MDRAT filtered damage signature for a T-joint embedded with a 70-mm horizontal delamination, subjected to loading at various angles. Figures 3.19 and 3.20 show the actual strain signatures, the DRAT filtered damage signature, and the MDRAT filtered damage signature of T-joints embedded with a 70-mm vertical delamination and a 20-mm delamination between the over-laminate and the filler section, respectively.

From Figures 3.18, 3.19, and 3.20, it can be seen that the modification made to the preprocessing algorithm (MDRAT) was reasonably effective in normalizing the strain signatures.

3.4.5.2   Anomalies Due to Crack Closure

Discrepancies were found to occur when the T-joint, embedded with an LH, SL, or LV delamination, was loaded at angles of 45° and 10° CCW to the y axis. This was because at these angles, the compressive nature of the force caused crack closure and hence increased the stochastic nature of the data for these two cases. The strain disturbance (along the bond line or surface) obtained because of the presence of a delamination, when crack closure occurred, was also found to be unique. Moreover, it also had a smaller magnitude (c and d in Figures 3.18–3.20) than that obtained when the crack tended to open.

3.4.5.3   ANN Test Results: Using MDRAT

The filtered DSTD was then used with the trained ANN to test the network's prediction and generalization accuracy. Figure 3.21 shows the normalized percentage prediction error obtained.

It was observed that the network was able to predict the presence and location of damage precisely in all the cases tested. Figure 3.21 shows the normalized percentage prediction error obtained by the network when predicting the extent of the delamination. It can also be seen from Figure 3.21 that the network predicts the extent of horizontal delaminations accurately.

However, unacceptably large errors were obtained for the vertical damage configurations. This can be explained with Figure 3.22. Figure 3.22 shows the MDRAT filtered strain distribution of T-joints embedded with LH, LV, and

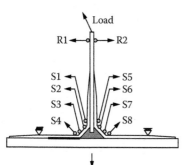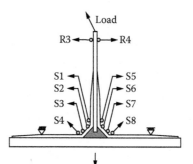

Damage Signature Test Database

	50 mm	75 mm		
Size	50 mm	75 mm	.	.
Location	LH	LH	.	.
Load	6000N	8000N	.	.
Load Angle	60 Deg	45 Deg	.	.
S1	−1.746	−1.449	.	.
S2	−3.239	−2.945	.	.
S3	−9.744	−10.358	.	.
S4	−9.889	−10.714	.	.
S5	13.968	15.292	.	.
S6	9.346	10.363	.	.
S7	3.935	4.744	.	.
S8	2.456	3.101	.	.
R1	−2.148	−2.205	.	.
R2	2.484	2.838	.	.
$X_D = \dfrac{R1}{Abs(R1)+Abs(R2)}$	−0.464	−0.437	.	.
Test Case	(a)	(b)	(c)	(d)

Undamaged Strain Signature

	5000N	5000N		
Load	5000N	5000N	.	.
Load Angle	60 Deg	45 Deg	.	.
$\varepsilon_1 \to$ S1	−1.417	−0.899	.	.
$\varepsilon_2 \to$ S2	−2.829	−1.936	.	.
S3	−7.494	−5.971	.	.
S4	−8.256	−6.716	.	.
S5	11.717	9.617	.	.
S6	7.347	6.111	.	.
S7	2.991	2.697	.	.
S8	2.04	1.932	.	.
R3	−1.79	−1.378	.	.
$\varepsilon_{10} \to$ R4	2.07	1.774	.	.
$X = \dfrac{R3}{Abs(R3)+Abs(R4)}$	−0.464	−0.437	.	.

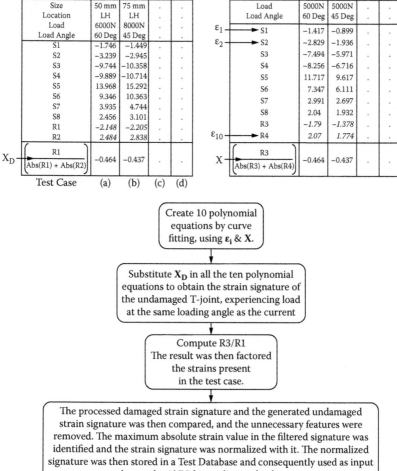

Create 10 polynomial equations by curve fitting, using ε_i & **X**.

Substitute X_D in all the ten polynomial equations to obtain the strain signature of the undamaged T-joint, experiencing load at the same loading angle as the current

Compute R3/R1
The result was then factored the strains present in the test case.

The processed damaged strain signature and the generated undamaged strain signature was then compared, and the unnecessary features were removed. The maximum absolute strain value in the filtered signature was identified and the strain signature was normalized with it. The normalized signature was then stored in a Test Database and consequently used as input data to the ANN for predicting the damage.

FIGURE 3.17
Schematic of the modification made to the DRAT algorithm to remove the effect of the load acting on the structure.

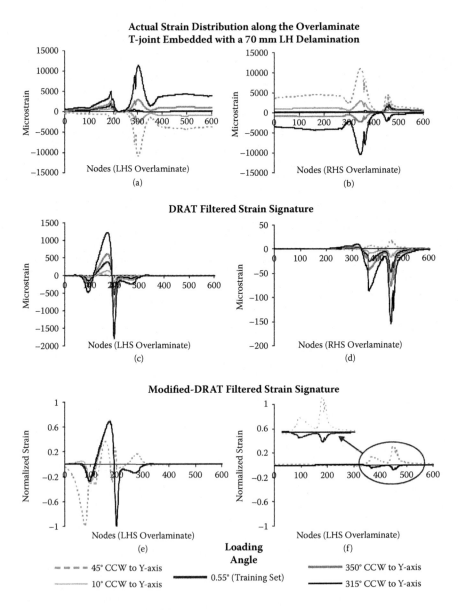

FIGURE 3.18
Comparison of the actual vs. the filtered strain signature for a T-joint embedded with a horizontal delamination, subjected to different loading angles.

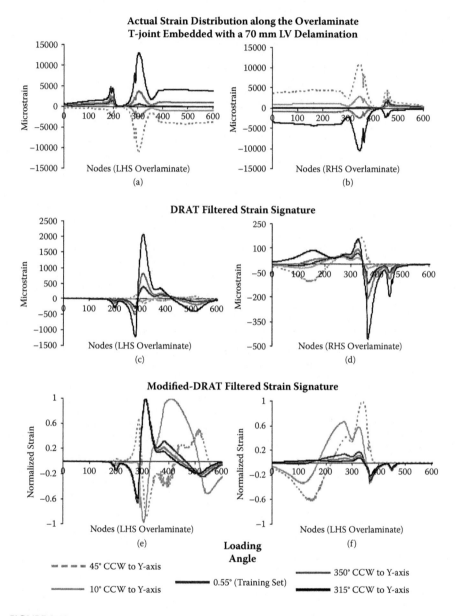

FIGURE 3.19
Comparison of the actual vs. the filtered strain signature for a T-joint embedded with a vertical delamination, subjected to different loading angles.

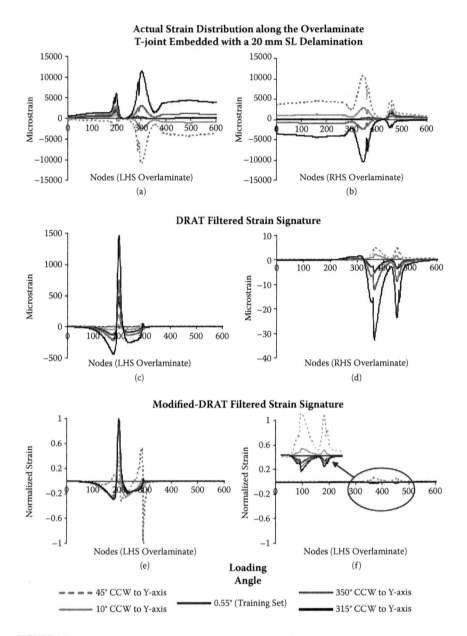

FIGURE 3.20
Comparison of the actual vs. the filtered strain signature for a T-joint embedded with an inclined delamination, subjected to different loading angles.

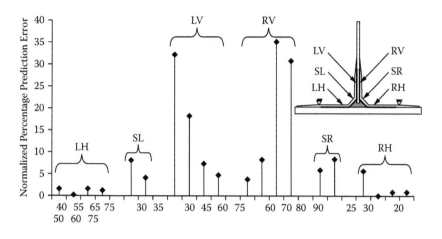

FIGURE 3.21
The percentage prediction error obtained when the trained network was used with the processed test set.

SL delaminations. From Figure 3.22 (a and b), it is apparent that the strain distribution along the RHS overlaminate of a T-joint, embedded with an LV delamination, subjected to different loading angles is not normalized as well as the strain distribution along the LHS overlaminate. This anomaly leads to a high prediction error in the case of LV delamination configuration. A similar sort of trend, with a lower magnitude, was also noticed for the SL delamination case. The presence of LV and SL delaminations affects not only the LHS strain distribution but also the RHS strain distribution, unlike the LH delamination. This effect increases in magnitude when the structure is loaded at an angle, which leads to the anomaly discussed.

3.4.6 Global Neural-Network Architecture for Incorporating Sequential Processing of Internal Sub Networks

To reduce the prediction errors produced, the training set is divided into two equal parts and trained separately by two neural networks. This is then combined into a virtual network by using the GNAISPIN. The architecture of the networks used is shown in Figure 3.23.

The network as shown in Figure 3.23 consists of two hidden layers with four and three neurons, respectively, and the output layer consists of four neurons (corresponding to LH, SLB, SLT, LV or RH, SRB, SRT, and RV damage locations, respectively). Tan-sigmoid activation function is used in the hidden layers and the log-sigmoid function in the output layer. The first network is trained with readings from sensors S1 to S4 (LHS of the bulkhead) and the second network is trained with the readings from sensors S5 to S8 (RHS of the bulkhead). The training set, consisting of strain signatures of T-joints embedded with different delamination sizes and configurations and

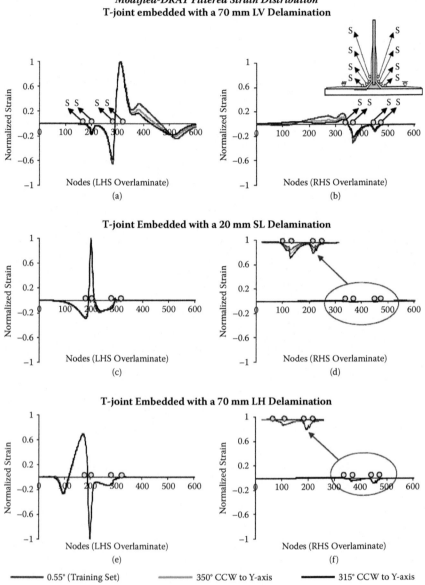

FIGURE 3.22
MDRAT filtered strain distribution of T-joints embedded with horizontal, vertical, and inclined delaminations, respectively, subjected to loading at different angles.

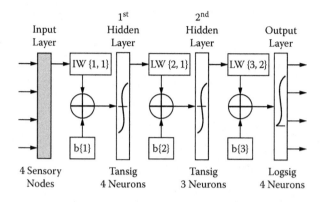

FIGURE 3.23
Architecture of the network used in tandem with the GNAISPIN method.

loaded at an angle of 0.55° CCW to the y axis, is filtered using the original DRAT algorithm. The filtered training set is then normalized with the maximum absolute strain value of the respective strain signatures. The normalized training set is then divided into two parts and used to train the two neural networks.

The GNAISPIN algorithm, programmed in MATLAB, is then used to virtually combine the two networks. GNAISPIN on execution creates a front end (graphical user interface), which requests the user to enter the structural strain readings obtained from the strain gauges. A schematic of GNAISPIN is shown in Figure 3.24. On receiving all the strain values from the damage signature test set (created earlier), GNAISPIN sequentially calls upon one sub network at a time and checks for delaminations.

Once all the sub networks are invoked, through a complete cycle, the results (i.e., the location of the delamination and its size) are displayed. GNAISPIN also incorporates the modified-DRAT preprocessing protocol developed earlier. The prediction result obtained using the GNAISPIN technique is presented in Table 3.2.

From Table 3.2, it is evident that using two separate networks, by the GNAISPIN technique, to train and test the networks produces much better results than the single-network case. The network locates the delamination accurately in all the cases shown in Table 3.2. The average damage size prediction accuracy was also found to be 94.5%. The improvement as a consequence of using GNAISPIN has resulted in an accuracy augmentation by up to 82.8%. This is by comparing the RV case for the 80-mm delamination in Figures 3.25a and 3.14b. Hence, it can be concluded that the modification made to DRAT in tandem with the GNAISPIN technique enhances the generalization ability of the ANN. Moreover, they also make the SHM system independent of the loading magnitude and loading angle acting on the structure.

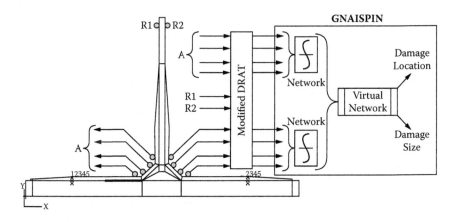

FIGURE 3.24
Schematic of the GNAISPIN incorporating the MDRAT algorithm.

The results obtained on testing the network with and without the GNAISPIN technique is shown graphically in Figure 3.25. The final step was to test and, if necessary, improve the performance of the SHM system in multiple-damage scenarios.

3.4.6.1   Multiple Damage Configuration Detection

Because the SHM system developed proved to be accurate in predicting the location and extent of delamination in a T-joint, regardless of the magnitude and angle of the load acting on the structure, the final step was to test the performance of the SHM system in multiple-damage scenarios.

The presence of a delamination in the T-joint structure not only affects the strain distribution above the delamination but also the strain distribution in the vicinity of the delamination. From Figure 3.26, it is clear that presence of a 70-mm LHS horizontal delamination affects the horizontal region of the overlaminate above the delamination, the inclined region of the overlaminate, and also causes minor disturbances in the RHS overlaminate region. A similar trend is also observed for the other delamination configurations.

Because of this occurrence, the number of sensors had to be increased and the location of the sensors also had to be modified for the SHM system to be able to detect and locate the multiple delaminations embedded in the structure. A total of 11 different multiple damage configurations were tested; and it was determined that 12 sensors (plus two reference sensors) were sufficient to predict the presence, location, and extent of the delaminations.

TABLE 3.2

Comparison of the Actual and Network-Predicted Size and Location of
Delaminations Embedded in T-Joints Subjected to Loading at Various Angles

Load	Load Angle	Location of the Delamination		Size of the Delamination			Normalized Error
		Actual	Predicted	Actual	Predicted	Difference	
N	Deg	Actual	Predicted	mm	mm	mm	%
5500	240	LH	LH	40	39.7	0.3	0.3
5000	320	LH	LH	55	55.5	−0.5	0.5
3500	300	LH	LH	65	64.5	0.5	0.5
3000	270	LH	LH	75	76.5	−1.5	1.5
4000	245	LV	LV	30	46.1	−16.1	16.1
7000	230	LV	LV	45	51.7	−6.7	6.7
5000	320	LV	LV	60	56.5	3.5	3.5
2500	300	LV	LV	75	62	13	13
4000	300	SL	SL	30	40.4	−10.4	10.4
6000	350	SL	SL	35	42.4	−7.4	7.4
4000	75	SR	SR	25	36.1	−11.1	11.1
3000	150	SR	SR	30	46.7	−16.7	16.7
4500	2	RV	RV	60	61.4	−1.4	1.4
6000	40	RV	RV	70	75.5	−5.5	5.5
5000	45	RV	RV	80	85	−5	5
2500	10	RV	RV	90	92.9	−2.9	2.9
3000	130	RH	RH	20	16	4	4
4500	45	RH	RH	50	47.7	2.3	2.3
5000	10	RH	RH	60	59	1	1
6500	0	RH	RH	75	75.1	−0.1	0.1
						Average error	**5.50%**

These sensors were positioned in such a way that it picked up the strain disturbance in the same plane as the delamination only. The position of the 12 sensors and the 2 reference sensors are shown in Figure 3.27.

To detect multiple delaminations, the network architecture had to be modified. Four networks were used, which were then combined together using the GNAISPIN technique to form a global network. The networks consisted of four layers; three hidden layers with three, four, and two neurons, respectively, and the output layer with one neuron. The training algorithm used was the resilient back-propagation algorithm. The activation function used

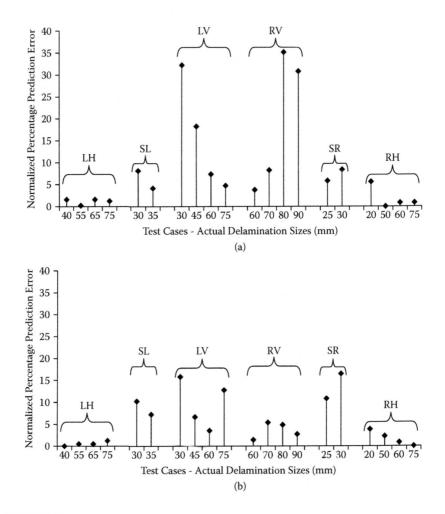

FIGURE 3.25
Network results obtained (a) without the use of GNAISPIN and (b) with GNAISPIN.

for the first and the second layer was tan-sigmoid, and log-sigmoid activation function was used for the third layer and the output layer.

The normalized training set, previously created, consisting of strain signatures of T-joints embedded with delaminations at various positions and subjected to a load of 5 kN at 0.55° CCW to the *y* axis, was divided into four parts and were used to train the respective networks. A separate test set was then modeled, which consisted of multiple delaminations embedded in the T-joint, subjected to loading at different magnitudes and also a few at different loading angles. The strain signatures obtained from these models were stored in the DSTD. The multiple delamination configurations studied are shown in Figure 3.28.

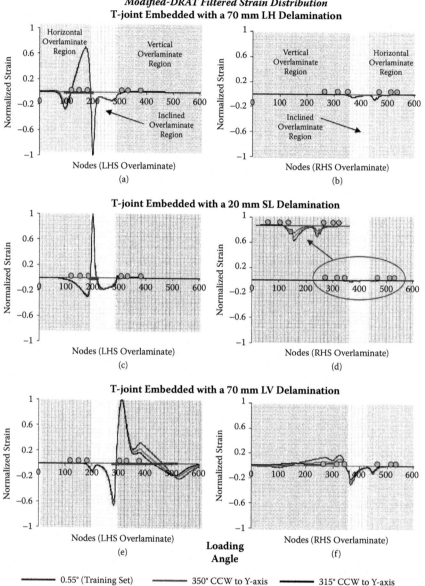

FIGURE 3.26
Strain disturbance caused by delaminations embedded in a T-joint subjected to loading at various angles.

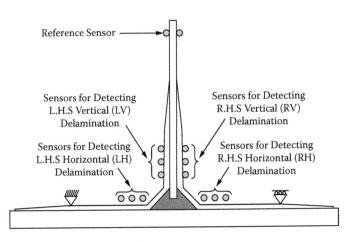

FIGURE 3.27
Sensor configuration for detecting multiple delaminations.

The strain signatures obtained from the DSTD were then filtered and normalized by using the MDRAT and the GNAISPIN techniques. The result obtained is presented in Table 3.3. Table 3.3 shows the actual vs. network-predicted size and locations of multiple delaminations embedded in T-joints, subjected to loading at various magnitudes.

The normalized prediction error in the table is represented as:

$$\text{normalized percentage error} = \left(\frac{\text{actual size} - \text{predicted size}}{\text{max. delamination size}} \right) 100$$

where the maximum length of the delamination is taken as 100 mm.

It is evident from Table 3.3 that the SHM system developed detected the presence of the delaminations precisely in all the test cases and also predicted the size of these delaminations with an average accuracy of 94.1%. However, a few test cases were noticed with prediction errors above 10%; this is due to excessive influence of the strain disturbance of one delamination over another in the damage signature. It was also noticed that the network failed to predict the delaminations when the delamination tended to close due to the loading angle. This error was because the training set consisted of strain signatures obtained from T-joints, embedded with delaminations, subjected to a load (5 kN) at an angle of 0.55° CCW to the *y* axis, which tended to open the delamination in all the damage configurations.

From the above study, it can thus be concluded that the SHM system developed consisting of the MDRAT, GNAISPIN, and the ANNs has proved to be capable of detecting the presence, predicting the location and extent of

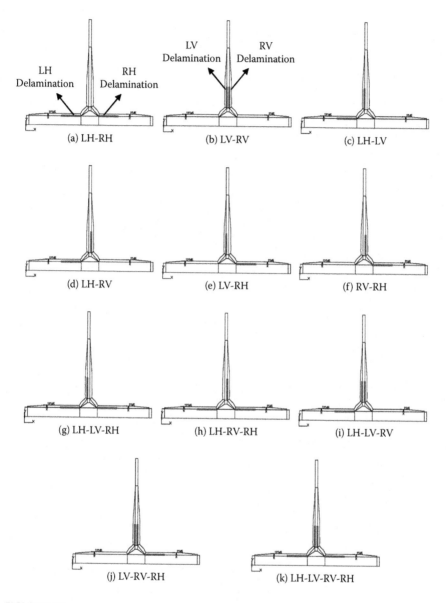

FIGURE 3.28
(a–k) Multiple damage configurations modeled for the test set.

multiple delaminations in the T-joint structure, regardless of the magnitude of the load applied and the loading angle.

TABLE 3.3

Comparison of the Actual and Network-Predicted Size and Location of Multiple Delaminations Embedded in T-Joints Subjected to Different Loading Magnitudes

Crack Configuration	Load N	LH Delamination A (mm)	P (mm)	NE (%)	LV Delamination A (mm)	P (mm)	NE (%)	RV Delamination A (mm)	P (mm)	NE (%)	RH Delamination A (mm)	P (mm)	NE (%)
LH-RH	5000	70	72.1	2.1							30	35.9	5.9
	5000	25	23.6	1.4							65	63.7	1.3
LV-RV	3500				100	109.1	9.1	100	117.1	17.1			
	5900				15	15.2	0.2	100	103.7	3.7			
LH-LV	6000	70	72.6	2.6	100	106.9	6.9						
	7000	30	30.5	0.5	60	56.2	3.8						
LH-RV	6000	35	38.8	3.8				75	80.1	5.1			
	7500	25	22.6	2.4				60	64.6	4.6			
LV-RH	8000				100	107.5	7.5				80	72.3	7.7
	5300				75	79.3	4.3				25	24.8	0.2
RV-RH	5000							60	67	7	31	39.1	8.1
	5000							100	108.9	8.9	26	26.2	0.2
LH-LV-RH	4520	35	42.8	7.8	100	107.7	7.7				25	24.7	0.3
	5560	55	59.6	4.6	20	21.4	1.4				63	59.8	3.2
LH-RV-RH	6750	24	23.2	0.8				50	60.4	10.4	55	52.9	2.1
	5550	58	61.6	3.6				15	19.5	4.5	58	55.6	2.4
LH-LV-RV	5950	70	74.1	4.1	100	104	4	100	110	10			
	5000	63	66.3	3.3	95	105.9	11	90	112.2	22.2			

		A	P	NE	A	P	NE	A	P	NE	A	P	NE
LV-RV-RH	4000				90	105.5	16	90	107.5	17.5	58	55.7	2.3
	3250				35	31.3	3.7	40	47.8	7.8	60	57.6	2.4
LH-LV-RV-RH	5000	67	73.8	6.8	25	24.1	0.9	35	23.4	11.6	55	51.8	3.2
	6500	15	23.1	8.1	30	33.6	3.6	40	46.4	6.4	24	22.8	1.2
LH-LV / 325°	5500	40	44.7	4.7	50	42.2	7.8						
LV-RH / 350°	6000				75	44.7	30				25	32.5	7.5
RV-RH / 90°	3000							40	44.5	4.5	40	40	0
RV-RH / 45°	3250							30	13.5	16.5	60	60	0
											Average error		5.9%

A, actual delamination; P, predicted delamination; NE, normalized.

3.5 Conclusions

3.5.1 Optimized ANNs

A comprehensive study was conducted on the various types of ANNs and training algorithms. The ANN played a key role in the SHM system developed. This was mainly because of its ability to establish mapping relationships between measurable features of structural damage and their physical parameters. Of the various types of networks, the feed-forward back-propagation neural network used along with either the resilient back-propagation training algorithm or the scaled conjugate gradient training algorithm appeared to produce the best results for the solution domain investigated in this study. An optimization method called the threefold cross-validation technique was used to choose the number of neurons in the hidden layers.

3.5.2 Damage Relativity Assessment Technique

A preprocessing tool, the DRAT, was developed to preprocess the strain signatures, stored in the DSD, before training the network. The filtered DSD was used to train a four-layer neural network. A separate test set, which was not a part of the training set, was then created to test the generalization performance of the trained network. It was determined that in using the SHM system developed (consisting of the ANN and the DRAT), 11 sensors placed on the surface of the T-joint structure were sufficient to determine the presence, predict the location (along the bond line), and estimate the extent of a horizontal delamination with an average prediction accuracy of 99.7%.

3.5.3 MDRAT and GNAISPIN Development

A training set consisting of the strain signatures of the T-joints, subjected to a constant loading magnitude and loading angle, embedded with delaminations at all the critical locations was created with the help a Patran Command Language (PCL) program. Three test sets were also created:

1. Varying the magnitude of loading.
2. Varying the angle of loading and the loading magnitude.
3. Varying the angle of loading, loading magnitude, and embedded with multiple delaminations.

However, in all the test cases, the size of the delamination differed from those used in the training set. The DRAT had to be modified to remove the effect of the load acting on the T-joint structure. Another program, GNAISPIN, was also written to virtually combine multiple ANNs and enhance the performance of the SHM system when detecting delaminations in structures experiencing variable loading angles or for detecting multiple delaminations embedded in composite structures.

3.5.4 The Universal ANN Incorporating MDRAT and GNAISPIN

It was determined that in using the MDRAT, GNAISPIN, and the ANNs, eight sensors (plus two reference sensors) were sufficient to detect the presence of delaminations and predict its location and extent with an average accuracy of 94.5%, regardless of the loading magnitude and loading angle. However, 12 sensors (plus two reference sensors) were needed to achieve an average accuracy of 94.1% in multiple-damage scenarios. From the research study conducted, it can be concluded that regardless of the complexity of the structure, the SHM system developed is capable of detecting the presence of delamination damage, predicting the location of the damage, and estimating the extent of the damage, regardless of the loading magnitude and the loading angle of the structure and the number of delaminations present in the structure.

It is also evident from the above study that the MDRAT protocol and the GNAISPIN technique provide a level of robustness that obviates the need to use numerous sensors for adequate SHM. This thus increases the prospects of deploying such an embedded Intelligence system in real-world structures because it will be able to augment the information received from a finite and manageable array of structural sensors.

References

Altinkok, N., and Koker, R. 2006. Modelling of the prediction of tensile and density properties in particle reinforced metal matrix composites by using neural networks. *Materials and Design* 27:625–631.

Ativitavas, N., Pothisiri, T., and Fowler, T. J. 2006. Identification of fibre-reinforced plastic failure mechanisms from acoustic emission data using neural networks. *Journal of Composite Materials* 40:193–226.

Dayhoff, J. E. 1990. *Neural network architectures: an introduction.* 1st ed. New York: Van Nostrand Reinhold.

Fang, X., Luo, H., and Tang, J. 2005. Structural damage detection using neural network with learning rate improvement. *Computers and Structures* 83:2150–2161.

Graham, D., Maas, P., Donaldson, G. B., Carr, C. 2004. Impact damage detection in carbon fibre composites using HTS SQUIDs and neural networks. *NDT&E International* 37:565–570.

Hagan, M. T., Demuth, H. B., and Beale, M. H. 1995. *Neural network design.* 1st ed. Boston: PWS Publications.

Haykin, S. 1999. *Neural networks—a comprehensive foundation.* 2nd ed. Upper Saddle River, NJ.: Prentice Hall.

Ishak, S. I., Liu, G. R., and Shang, H. M. 2001. Locating and sizing of delamination in composite laminates using computational and experimental methods. *Composites Part B: Engineering* 32:287–298.

Jia, J., and Davalos, J. F. 2006. An artificial neural network for the fatigue study of bonded FRP–wood interfaces. *Composite Structures* 74:106–114.

Kesavan, A., Deivasigamani, M., John, S., and Herszberg, I. 2005. Detection of delaminations in composite structures using optimised neural networks. In: *Proceedings of ICAI-05,* Las Vegas, USA, June 27–30, 2005.

Kesavan, A. 2006. Embedded Intelligence in structural health monitoring using artificial neural networks. PhD thesis, RMIT University, Melbourne, Australia.

Kesavan, A., John, S., and Herszberg, I. 2008. A structural health monitoring system for polymeric composite structures using artificial neural networks. *Journal of Structural Health Monitoring* 7:203–213.

Lee, J. J., and Yun, C. B. 2006. Damage diagnosis of steel girder bridges using ambient vibration data. *Engineering Structures* 28:912–925.

Nyongesa, H., Otieno, A. W., and Rosin, P. L. 2001. Neural fuzzy analysis of delaminated composites from shearography imaging. *Composite Structures* 54:313–318.

Orazio, T. D., Guaragnella, C., Leo, M., and Spagnolo, P. 2005. Defect detection in aircraft composites by using a neural approach in the analysis of thermographic images. *NDT&E International* 38:665–673.

Seo, D. C., and Lee, J. J. 1999. Damage detection of CFRP laminates using electrical resistance measurement and neural network. *Composite Structures* 47:525–530.

Simpson, P. K. 1990. *Artificial neural systems: foundations, paradigms, applications, and implementations.* 1st ed. New York: Pergamon.

Su, Z., and Ye, L. 2004. Lamb wave-based quantitative identification of delamination in CF/EP composite structures using artificial neural algorithm. *Composite Structures* 66:627–637.

Su, Z., and Ye, L. 2005. Digital damage fingerprints and its application in quantitative damage identification. *Composite Structures* 67:197–204.

The Mathworks. 1999. *Neural network toolbox-user's guide.* Mathworks, 10 September 2003. http://www.mathworks.com/access/helpdesk/help/toolbox/nnet/.

Xu, Y. G., Liu, G. R., Wu, Z. P., and Huang, X. M. 2001. Adaptive multilayer perceptron networks for detection of cracks in anisotropic laminated plates. *International Journal of Solids and Structures* 38:5625–5645.

Yam, L. H., Yan, Y. J., and Jiang, J. S. 2003. Vibration-based damage detection for composite structures using wavelet transform and neural network identification. *Composite Structures* 60:403–412.

Yan, Y. J., Cheng, L., Wu, Z. Y., and Yam, L. H. 2007. Development in vibration-based structural damage detection technique. *Journal of Mechanical Systems and Signal Processing* 21:2198–2211.

4

Damage Identification and Localization of Carbon Fiber-Reinforced Plastic Composite Plate Using Outlier Analysis and Multilayer Perceptron Neural Network

F. Mustapha, S. M. Sapuan, K. Worden, and G. Manson

CONTENTS

ABSTRACT   This chapter presents a novel approach to detecting and localizing structural defects based on a novelty detection method called outlier analysis and a multilayer perceptron (MLP) neural network. The main specimen used here was a rectangular carbon fiber-reinforced plastic composite plate. The effectiveness of the selected approach was assessed by analyzing the experimental data acquired from the PZT patches. The scope of this present work also comprises an investigation of the scattering effect of an ultrasonic-guided wave on the tested plate under both damaged and undamaged conditions. The wave propagation is sequentially transmitted and captured by eight PZT patches bonded on the plate, forming a sensor network on the tested composite's rectangular structure. An in-house eight-channel multiplexer is incorporated into this small-scale and low-cost structural health monitoring system to effectively swap the PZTs role from sensor to actuator and vice versa. The Real Time Damage Demonstrator software is primarily developed to acquire and store the waveform responses. These sets of scattering waveform responses representing normal and damaged conditions are transformed into a set of novelty indices that ultimately determine the true conditions of the tested structure. The acquired novelty indices representing the available sensor paths are used as the inputs for the neural network incorporating the multilayer perceptron architecture to compute and predict the damage location as the x and y location on the tested composite plate.

4.1   Novelty Detection

This is a two-class problem which has the advantage that unsupervised learning can be used. The principle of this approach is to establish a description of normality using features representing the undamaged condition of the structure and then test for abnormality or novelty when new test data become available. Tarassenko et al. (1995) defined the principle of novelty detection as offering an approach to the problem of fault detection, which only requires the normal class to be defined. They also noted that the key issue in novelty detection is the development of appropriate models of normality. If a statistical model of normality in feature space is to be employed, then they proposed that the complexity of the model must match the amount of the training data available.

Numerous works cited by Markou and Singh (2003) highlight the applications of novelty detection in various fields such as fault detection, radar target detection, detection of masses in mammograms, hand-written digit recognition, Internet and e-commerce, statistical process control, and several others. Several important issues that relate to novelty detection such as the principle of robustness and tradeoff, uniformity of data scaling, parameter minimization, generalization, independence, adaptability, and computational complexity can also be found in this paper. They have also reviewed several statistical approaches to novelty detection (such as parametric and nonparametric approaches). At the end of the paper, they have highlighted the advantages such as the cheapness to compute and straight-forwardness, which can be achieved when adapting novelty detection using the statistical approaches.

More evidence about the successfulness of novelty detection methods can be seen in Manson et al. (2003); here they employed a novelty detection method based on measured transmissibilities from a simplified model of a metallic aircraft wing box. Three types of novelty detection algorithm were discussed in this paper: outlier analysis (OA) (Worden et al., 2007a), probability density estimation Bishop (1994), and artificial neural networks (Worden, 1997; Sohn et al., 2001). For the current work, the former method will be the governing novelty detection tool employed in this chapter, and the theoretical background is explained next.

4.2   Outlier Analysis

According to Markou and Singh (2003), the statistical approaches to novelty detection are mostly based on modeling data generated from their statistical properties. They have also stressed that, by using this information (such as the mean and standard deviation), the prediction of whether a test sample comes from the same distribution or not can be made. On the other hand, the development of the statistical approach to damage detection is mainly concerned with the implementation of the algorithms that operate on extracted features to determine the damage state of the structure (Sohn et al., 2001). If the observed features indicate a significant change from the description of normality, the data will be flagged as an outlier or indicated as belonging to a damaged state. Analysis of outliers is even applicable when data from the damaged structure are not available for comparison. The use of OA for damage detection and its success has been extensively reported in Worden et al. (2000a), Worden et al. (2000b), and Sohn et al. (2002); a detailed technical description is highlighted in the next section.

4.2.1   Outliers in Univariate Data

$$z\zeta = \frac{\left| x\zeta - \bar{x} \right|}{s}$$

(4.1)

Equation (4.1) represents an outlier detector for univariate data, where $x\zeta$ is the measurement corresponding to a potential outlier, and $\bar{x}$ and s are the mean and standard deviation of a normal undamaged sample, respectively. An outlier in a data set is an observation that is surprisingly different from the rest of the data in some sense and therefore is thought to be generated by a different mechanism to the other data. The discordancy of a candidate outlier is some measure that can be compared against a corresponding objective criterion and allows the outlier to be judged as being statistically likely or unlikely to have come from the assumed generating model. The application to damage detection is clear; the discordancy should be evaluated with respect to a model constructed from an undamaged condition of the system of interest. Barnett and Lewis (1994) is the standard reference for OA.

4.2.2   Outliers in Multivariate Data

$$D\zeta = (\{x\zeta\} - \{\bar{x}\})_T \,[S]{-1}(\{x\zeta\} - \{\bar{x}\}) \tag{4.2}$$

A multivariate data set consisting of n observations in p variables may be represented as n points in a p-dimensional vector space. It is clear that outlier detection is more difficult than in the univariate situation due to the potential outlier having the ability to appear more hidden in the data mass. That said, many of the ideas and techniques associated with the detection of outliers in multivariate data follow on from those applicable to univariate problems. The discordancy test, which is the multivariate equivalent of equation (4.1), is the Mahalanobis squared distance measure as given in equation (4.2), where $\{x\zeta\}$ is the potential outlier datum, $\{x\}$ is the mean vector of the sample observations, and $[S]$ is the sample covariance matrix for normal condition data. T indicates a matrix transpose.

In many practical situations, the outlier is not known beforehand and so the test would need to be conducted inclusively. In structural health monitoring case studies however, the potential outlier is always known beforehand, and so it is more sensible to calculate a value for the Mahalanobis squared distance without this observation "contaminating" the statistics of the undamaged data. Whichever method is used, the Mahalanobis squared distance of the potential outlier is checked against an appropriate threshold value, as in the univariate case, and its status determined.

Determination of the appropriate threshold is critical. This value is dependent on both the number of observations and the number of dimensions of the problem being studied Worden et al. (2000b), A Monte Carlo method based on extreme value statistics was used here to arrive at the threshold value. Worden et al. (2000b) explain how to calculate this value or the critical value of discordancy.

4.2.3 Neural Network Approaches

In recent years, various techniques based on neural networks have been applied to the problem of structural damage detection (Worden et al., 1993; Worden, 1991). References collated in Zapico et al. (2001) and Tarassenko (1998) show how the multilayer perceptron (MLP) neural network can be applied to many problems with a dependence on the knowledge of a problem domain to indicate how the design and training of the neural network is carried out. In the conventional method for structural damage identification using Lamb wave propagation, the parameters known as time of flight, peak-to-peak values, *fd* frequency thickness product, complex multimode waveform responses, and the development of complicated dispersion curves are of a high priority to be investigated before any diagnosis regarding the state of the monitored structure can be made (Alleyne, 1992). A statistical approach employing OA can be seen as an alternative approach in alleviating the above difficulties when adapting the elastic wave as a means of damage monitoring. In addition, OA has also proved to be a fast and simple solution for detecting structural degradation (Worden et al., 2000a).

According to Rytter (1993), damage identification problems can be categorized into a hierarchical structure, namely, detection, localization, assessment, and prediction. Because novelty detectors can distinguish the two states of structural condition, the detection level is achieved. This present work aims at achieving the second stage of the hierarchical structure, that is, localization, by employing a neural network algorithm using an MLP. Fundamentally, the general idea of the neural network is to represent a nonlinear functional mapping between a set of input variables into a set of output variables. The philosophy of the approach is simply to establish the secondary features of the novelty index values computed from the primary features of the novelty detection procedures. To express the current work in terms of this neural network terminology, the acquired novelty indices representing the available sensor paths will be used as the inputs to the network and the location of the damage will be the output of the network.

According to Zapico et al. (2001), the success of the neural network is very dependent on incorporating the knowledge of the problem domain into the design and training of the neural network. To achieve this, the first phase of the neural network application to the damage localization problem is to find the best architecture or neural network topology in predicting the damage location as x and y coordinates on the isotropic plate (best generalization performance). Overall, there are four main tasks involved in this process. These are the partitioning of the acquired novelty indices for each damage position in the ratio of 40:20:20 for the training, validation, and the testing phase, respectively (overall, there are 80 sets of novelty indices, also known as a cross-validation set); training using the MLP model; selecting the optimal network; and finally testing the trained and the best validated network on the remaining test set.

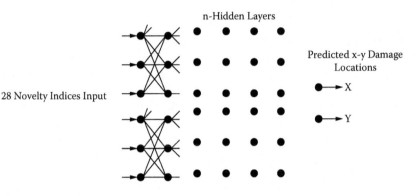

FIGURE 4.1
MLP architecture for damage detection and localization problems.

4.2.4 MLP Neural Networks

The MLP is one of the most commonly used neural network architectures and uses a multilayer feed-forward structure. The MLP consists of a collection of connected nodes, namely, the input layer nodes, hidden layer nodes, and finally the output nodes. Worden (1996) discusses the subject in some detail. In this present work, the input nodes correspond to the novelty indices data sets, hidden layers are used for computation, and the outputs are the prediction of the coordinates of the damage location in the x and y directions. The MLP architecture for this problem domain is illustrated in Figure 4.1.

The MLP neural network is a set of nodes arranged in a manner of layers. The initiation of the above network architecture begins when a set of signal (novelty indices) values pass into the input layers (28 paths equivalent to 28 nodes), progress forward through the hidden layers, and the results (predicted damage locations) finally emerge at the output layer (prediction of the damage locations). Each node i is connected to each node j in the preceding and the following layers through a connection of weight w_{ij}. Signals pass through each node as follows: in layer k (hidden layer), a weighted sum is performed at each node i of all the signals $x_j^{(k-1)}$ from the preceding layer $k-1$, giving the excitation $z_i^{(k)}$ of the node; this is then passed through a nonlinear activation function f to emerge as the output values $x_i^{(k)}$ to the next layer, which is highlighted in equation (4.3) below:

$$x_i^{(k)} = f^{\left(z_i^{(k)}\right)=f\left(\sum_j w_{ij}^{(k)} x_j^{(k-1)}\right)} \tag{4.3}$$

The activation function f in this case is restricted to $f(x) = \tanh(x)$. One node of the network, the bias node b, is special in that it is connected to all other nodes in the hidden and output layers; the output of the bias node is held fixed throughout to allow constant offsets in the excitations z_i of each node.

Thus, equation (4.3) is transformed into:

$$x_i^{(k)} = f^{\,(z_i^{(k)})=f\left(\sum_j w_{ij}^{(k)} x_j^{(k-1)} + b_j^{(k-1)}\right)} \tag{4.4}$$

Finally, the output-unit activation function will generate the output values y_k. Here the consideration of the three forms of activation functions should be given.

For regression problems, an appropriate choice is a linear function of the form;

$$y_k = x_j^{(k)} \tag{4.5}$$

For classification problems involving multiple independent attributes, logistic sigmoidal activation functions can be applied to each of the outputs independently, so that:

$$y_k = \frac{1}{1 + \exp(-x_j^{(k)})} \tag{4.6}$$

For the more usual kind of classification problem in which one has a set of c mutually exclusively classes, one uses the softmax activation function of the form:

$$y_k = \frac{\exp(x_j^{(k)})}{\sum_{j'} \exp(x_{j'}^{(k)})} \tag{4.7}$$

For this work only the first and the third forms of the activation functions were taken into consideration. The main reason for selecting the softmax activation function in deriving solution toward classification approach is that it is a multiclass problem and the outputs can be interpreted as probabilities of class membership, conditioned on the outputs of the hidden units (Bishop, 1995).

4.3 Methodology

The flowchart in Figure 4.2 emphasizes the methodology of the adopted neural network computations to accomplish the best possible damage location on the tested isotropic plate. The freeware software Netlab (2005) is the main tool implemented for this MLP neural network application. All the scripts are written in Matlab and are compatible with any Matlab platform (in this case Matlab version 6.5 is used). The basics of Netlab as well as its theoretical description can be found in Nabney (2002). For each neural network computation, the input data were preprocessed by scaling them into a range of [0 1] or [–1 1].

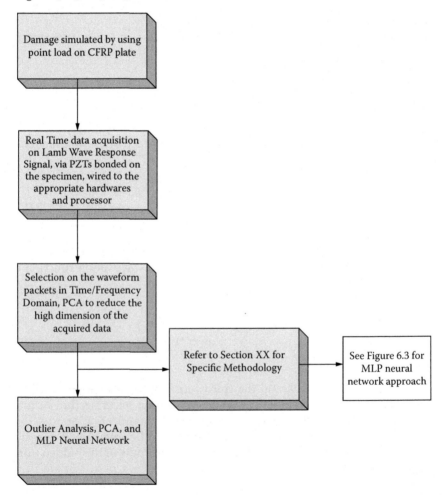

FIGURE 4.2
Generic methodology.

For the regression problem, after the partitioned data have undergone the training phase, the mean and the median of the predicted vectors of the coordinate axis are extracted and plotted. For a better analysis, the percentage of nondimensional mean square error (MSE) is plotted to evaluate the performance of each of the tested hidden layer units used in predicting the damage locations (see Results section). Equation (4.8) gives the percentage of nondimensional MSE. On the other hand, for the classification problem, the technique of locating the predicted damage is done by assessing which network topology produces the best classification rate, together with the generation of a summary result called a confusion matrix.

$$\text{nondimensional MSE} = \frac{\sum_{i=1}^{N} \pi r_i^2}{\text{area of plate} \times N} \times 100\% \tag{3.8}$$

where $r_i^2 = (x_{ti} - x_{pi})^2 + (y_{ti} - y_{pi})^2$,

x_t, y_t are the coordinates for the targeted damage location.
x_p, y_p are the coordinates for the prediction damage location.
N is the number of targeted/predicted damage points.

In both the regression and classification problems, to enhance the efficiency of the adapted MLP algorithm, weight decay or regularization is added to the MLP methodology as highlighted in the flowchart in Figure 4.3. One may refer to Bishop (1995) for a better understanding of weight decay or regularization problems in neural network. According to this reference, the technique of regularization encourages smoother network mapping by adding a penalty factor to the error function in the neural network computation. One of the simplest forms of this penalty factor is termed *weight decay*. This method offers significant improvements in the network generalization (Bishop, 1995).

4.4   Experiment Setup and Data Handling Procedures

The experimental setup for launching the ultrasonic-guided wave in the selected specimen and the overall layout of the structural health monitoring system architecture for this work is illustrated in Figure 4.4. The system is divided again into three main subsystems, namely,

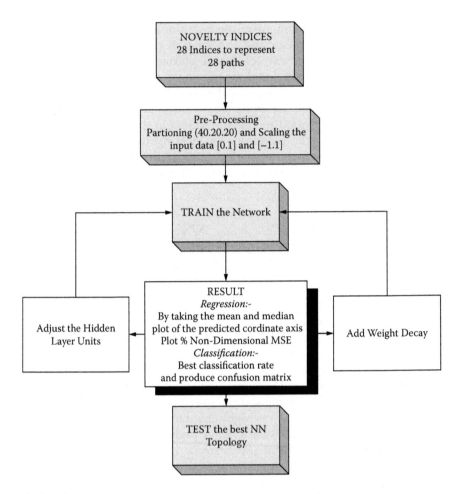

FIGURE 4.3
Methodology adopted for MLP neural network approach.

   i. The specimen geometry,

  ii. The hardware configuration,

 iii. The software implementation.

4.4.1　Specimen Geometry

The length, width, and thickness of the plate are 305, 215, and 2 mm, respectively. The plate is made of carbon fiber–reinforced plastic (CFRP) shaped in a rectangular form. The fiber orientation for this plate was unidirectional with zero degree orientation [UD] 0. The plate has nine PZT sensors, and the sensor configuration is illustrated in Figure 4.5. The simulated locations

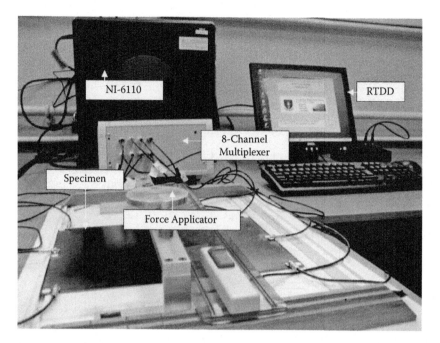

FIGURE 4.4
Structural health monitoring overall layout.

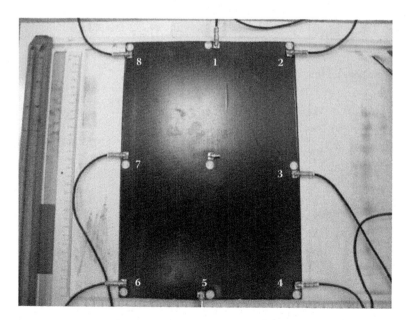

FIGURE 4.5
Sensor configurations.

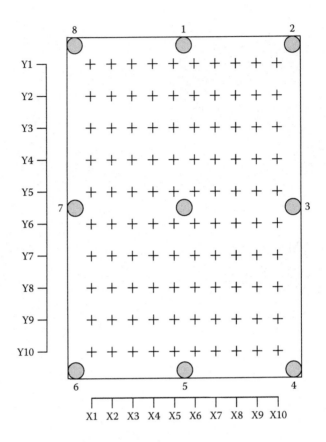

FIGURE 4.6
Damage locations for time domain analysis.

TABLE 4.1

Coordinate axis for damage locations

Item	X coordinates	Y coordinates
X1,Y1	20	290
X2,Y2	40	260
X3,Y3	60	230
X4,Y4	80	200
X5,Y5	100	170
X6,Y6	120	140
X7,Y7	140	110
X8,Y8	160	80
X9,Y9	180	50
X10,Y10	200	20

of damage are represented by the 100 points shown in Figure 4.6. Table 4.1 outlines the coordinates for the simulated damage for this plate.

4.4.2 Hardware Requirements

An in-house eight-channel multiplexer having BNC inputs and BNC 2110 as an output to the DAQ card series PCI NI-6110 was the main hardware used in this experiment. The processing engine for acquiring and transmitting the waveform signal was written in Matlab 6.5 using the DAQ Toolbox running under the Windows operating system.

4.4.3 Software Implementation

GUIDE in Matlab 6.5 is used in designing and developing the graphical user interface for the real-time damage demonstrator. The main objective of the developed real-time damage demonstrator is to act as a technical tool for the data acquisition and feature extraction processes.

4.4.4 Data Handling Procedure

Damage on the tested plate was simulated by a localized strain as a result of a force applicator applied on the surface of the plate, as demonstrated in Figure 4.6. There were 100 damage locations equally spaced by 20 and 30 mm in the x and y directions, respectively. The data acquisition for normal and damage conditions for the anisotropic plate again consisted of 100 sets of observations each of 2000 points or dimensions. This points/dimension would be resampled to 50 points or dimensions to suit the statistical technique employing OA.

4.4.5 Data Acquisition

Input signal as five cycle tone burst signals was employed here, and the procedure of transmitting and receiving waveform responses were done by inputting certain appropriate variables on the developed real-time damage demonstrator graphical user interface. A good signal-to-noise ratio waveform is required to proceed further with the OA. Typical waveform responses with reasonably good signal-to-noise ratio due to this excitation signal can be observed in Figures 4.7 and 4.8. Overall, there were 28 sensor paths involved in every data acquisition process and the pairs of sensors involved can be seen in Table 4.1. There were initially 100 observations of 2000 points or dimensions that later would be resampled to 50 points or dimensions. Once the number of the undamaged feature vectors was adequate for the novelty detection, the specimen was then switched to a damaged condition, simulated by placing an appropriate load on the force applicator. The data comprising 100 damage locations were to be

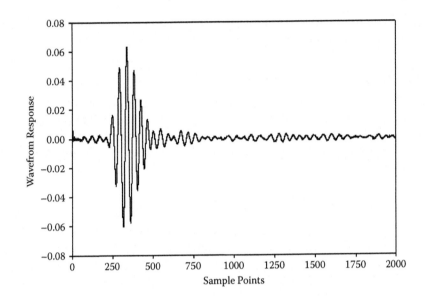

FIGURE 4.7
Typical waveform response for paths 4–5 (actuator no. 4, sensor no. 5).

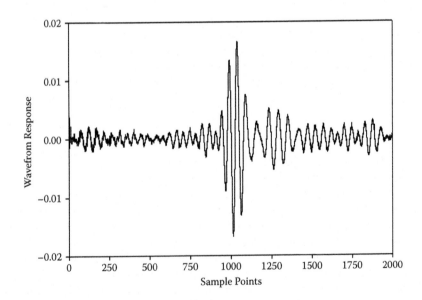

FIGURE 4.8
Typical waveform response for paths 2–4 (actuator no. 2, sensor no. 4).

investigated from position (X1,Y1) to (X10,Y10). Overall, there were 100 sets of damaged features comprising 80 observations by 28 paths (representing the sensor paths) and 50 dimensions and 100 observations with similar path sensors for the undamaged features stored for this work in detecting and localizing the damage.

4.4.6 Analyses of the Experimental Results

The analyses of the acquired results presented here considered damage in a simulated form, where only a point load was placed on the force applicator.

4.4.7 Feature Selections and Extraction

Figure 4.9 shows an example of the extracted or zoomed area of the wave response. As described earlier in a previous chapter, to construct a suitable mean vector $\{x\}$ and covariance matrix $[S]$ for the OA study, the extracted area comprised 250 points and was resampled to 50 points, to equate to 50 dimensions in outlier terminology.

4.4.8 Result of OA

Similarly to the previous chapter, by adapting equation (3.2), the novelty index (Mahalanobis squared distances) for all points in the normal and the damage data sets can be derived. The mean and covariance values in this

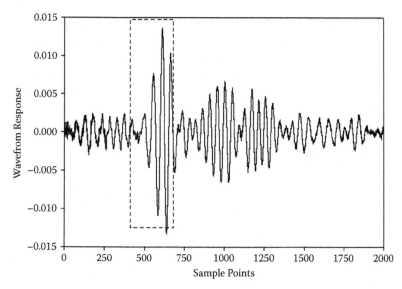

FIGURE 4.9
Waveform signal with the zoomed area (rectangular dotted line).

equation were again derived from the normal data sets. Figures 4.10 to 4.14 show the novelty index plot with respect to the sensor path configurations. The threshold value (---) represented by the dotted line was again 290.2 and 2.462 in linear and logarithmic scales, respectively, for the case of 50 dimensions and 100 observations. Similarly, to previous cases, all the novelty index plots are in logarithmic scales.

4.4.9   OA Result for Anisotropic Plate

The expectation of experiencing outlier behavior on the designated sensor paths that are in the vicinity of damage occurring on the anisotropic plate is still achievable in this work (see Figure 4.10 for damage 1). Unfortunately, not all of the sensor paths act accordingly to the simulated damage pattern. Unexpected paths that are flagged as outlier can be seen here as the damage has progressed further on the next points as illustrated in Figures 4.11 to 4.14. Based on these plots, there exist several unexplained behaviors of a few of the sensor paths. Paths 1–3 and 3–4 are the examples of the obvious paths that have the misbehaving patterns with respect to the damage region because they do not cross the damage point. Figure 4.15 clearly highlights how these paths react toward damage presented at locations 11 and 42.

Inspired by this peculiar phenomenon and the confidence of the OA in detecting damage, further investigation was carried out on why this unexplained condition kept on happening. To achieve this, a point load was applied at damage location 28 and by allowing only these two sensor paths to be active during the data acquisition process. The arrangement of the investigation process can be seen in Figure 4.16 (circular dotted line represents the damage location). Both of the structural conditions were tested on the selected sensor paths and the results reveal that, if the damage is present at the designated location (no. 28), both of the paths flagged as outlier; whereas for the nondamage condition, both of the MD values were below the threshold values. Thus, nonviolation on the process of detecting a damage using OA can be made here, as still damage is being detected when the defect is taking place on the plate structure. On the other hand, few assumptions can be made on why sensor path flagged as outlier. Nevertheless, the vagueness of using this approach alone in localizing the damage still persists and the MLP neural network will be included here to complete the required task.

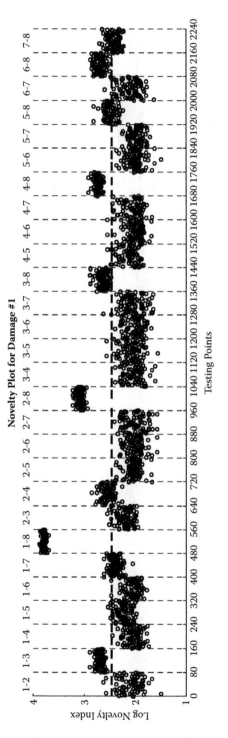

FIGURE 4.10
Novelty index plot for damage no. 1.

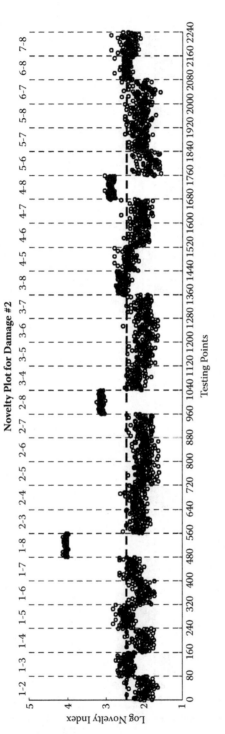

FIGURE 4.11
Novelty index plot for damage no. 2.

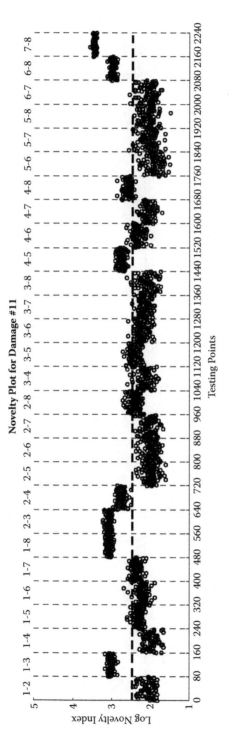

FIGURE 4.12
Novelty index plot for damage no. 11.

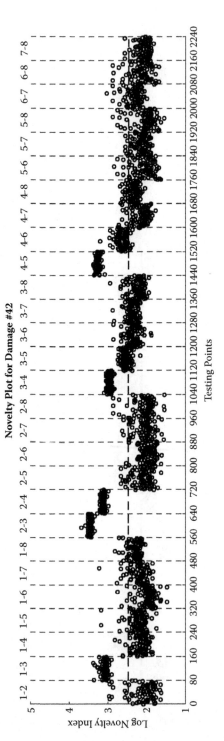

FIGURE 4.13
Novelty index plot for damage no. 42.

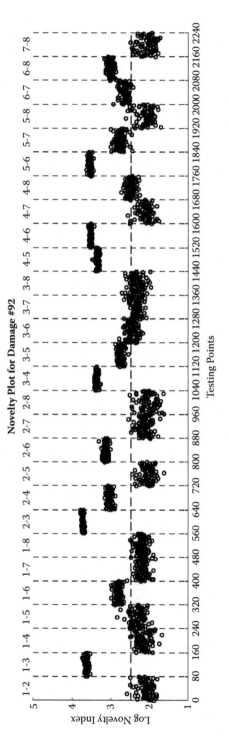

FIGURE 4.14
Novelty index plot for damage no. 92.

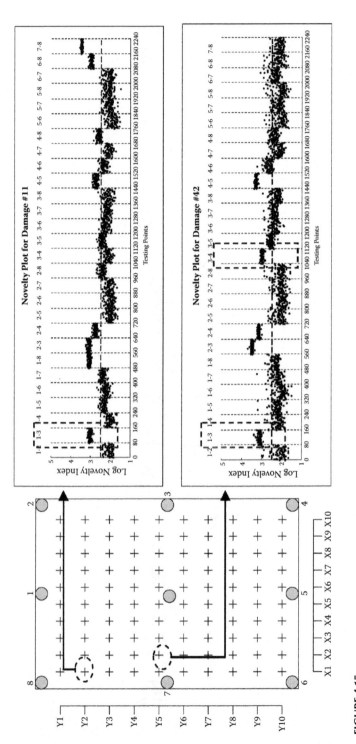

FIGURE 4.15
Paths 1–3 and 3–4 for being outlier.

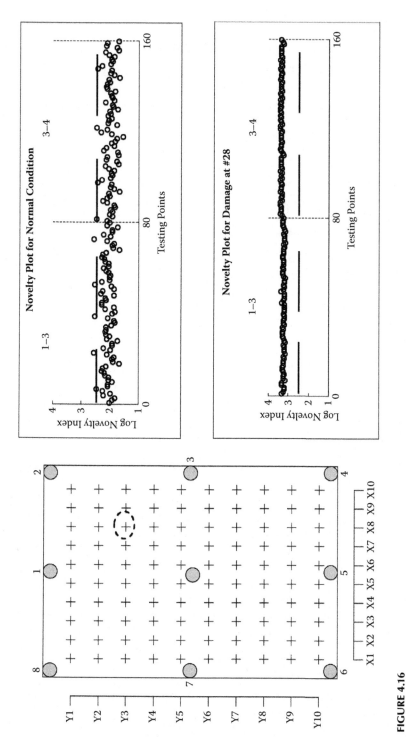

FIGURE 4.16

Test on normal and damage condition for paths 1–3 and 3–4.

4.5   Results of MLP Neural Network

Several methods introduced by Mustapha et al. (2005) have been adopted here in acquiring the best possible location for the simulated damage occurred on this anisotropic plate. Only the regression problem was included here in arriving to this solution.

4.6   Regression Problems

4.6.1   Time Domain Analysis

A nondimensional MSE plot for the prediction value is again plotted to better understand the methods for predicting the damage location (see Figure 4.19). Examples of the predicted damage locations for the case of 15 hidden layer units with input scaling in the range of [–1 1] and [0 1] can be seen in Figures 4.17 and 4.18, respectively.

4.6.2   MLP Neural Network for the Overall Plate Area

As for this work, to determine the damage location on the plate, various hidden layer units were trained together with the input scaling ranging between [–1 1] and [0 1]. The best prediction will be justified by the lowest MSE value (see Figure 4.19). No weight decay results are included as this parameter made the prediction worse.

Figures 4.17 and 4.18 demonstrate the MLP neural network computation for disseminating the damage localization results based on the similar number of hidden units (10) and input scaling in the range of [–1 1] and [0 1] in an orderly manner. For this approach, the best prediction occurs when the number of hidden layer units is equal to 10 with the input in the scale of [–1 1] (see MSE plot in Figure 4.19).

Generally, all the novelty index plots generated employing this approach still give good damage location prediction, although several of the sensor paths behaved unexpectedly (MSE values in the range of 2–7% error). This occurrence can probably be the uniqueness of the MLP neural network computation in accepting the sets of unique novelty index vectors as the input parameters.

4.6.3   Symmetrical Approach

The results on the OA are presented first and are depicted in Figures 4.21 to 4.25. The new sensor paths (for the symmetrical approach) were executed in this approach (see Figure 4.20 and Table 4.2). The results generated from

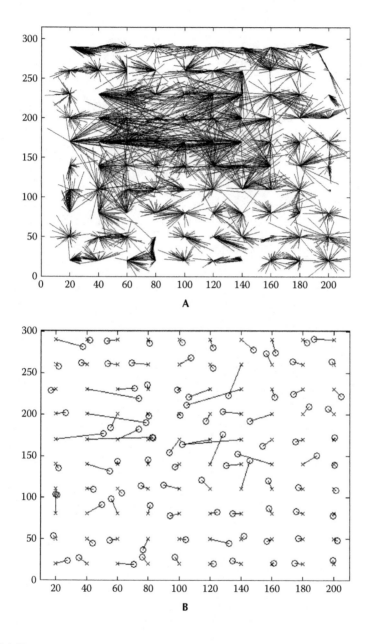

FIGURE 4.17
Hidden layer units = 15 [–1 1].

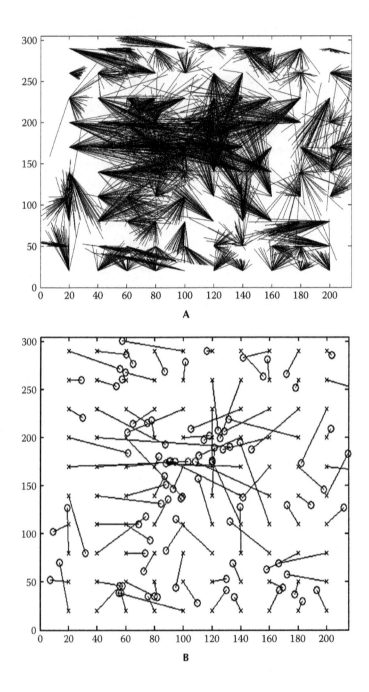

FIGURE 4.18
Hidden layer units = 15 with scaling [0 1].

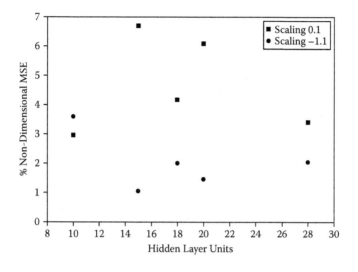

FIGURE 4.19
Nondimensional MSE plot for time domain analysis.

the MLP neural network computation are compiled in Figures 4.26 to 4.28. A hidden layer with 20 units is highlighted here with preprocessing input scaling in the range of [0 1] and [−1 1] (Figure 4.27) together with the best prediction damage location as shown in Figure 4.28. The MSE plot is depicted in Figure 4.29.

4.6.4 OA for the Symmetrical Approach

Unlike the OA results presented in the time domain approach discussed previously, almost all the novelty index plots from this smaller area of approach have proven to be viable in characterizing the behavior of the outlier with regard to the damage simulated on the anisotropic plate. Examples of these promising MD value plot results in evaluating the behavior of anisotropic plate property toward damage can be visualized from Figures 4.21 to 4.25.

4.6.5 MLP Neural Network for the Symmetrical Approach

Interestingly enough, the methodology adopted in the isotropic plate in acquiring the best prediction of damage location is also applicable here. The lowest MSE for this case again has the effect of regularization on the MLP neural network computation as illustrated in Figure 4.29. The value of the weight decay that produced the best approximation on the damage location was equal to 0.1 with hidden layer unit amount to 15 and in the range of [−1 1] for the input scaling (see Figure 4.28). Examples of other hidden layer units and various input scalings can best be seen from Figures 4.23 to 4.27.

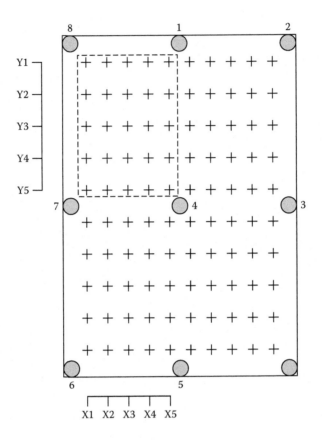

FIGURE 4.20
Damage locations for symmetrical approach.

TABLE 4.2

Coordinate axis for damage locations for symmetrical approach.

Item	X-Coordinates	Y-Coordinates
X1,Y1	20	290
X2,Y2	40	260
X3,Y3	60	230
X4,Y4	80	200
X5,Y5	100	170

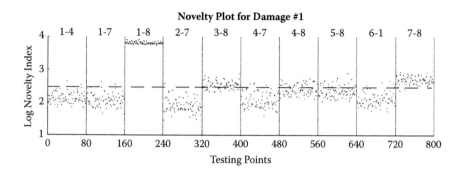

FIGURE 4.21
Novelty index plot for symmetrical approach of damage no. 1.

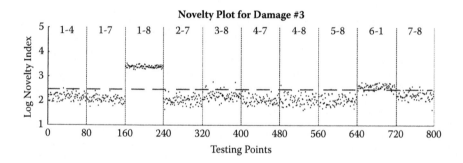

FIGURE 4.22
Novelty index plot for symmetrical approach of damage no. 3.

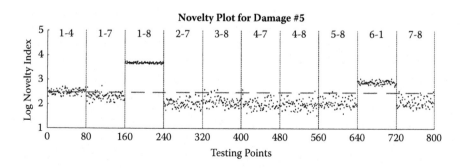

FIGURE 4.23
Novelty index plot for symmetrical approach of damage no. 5.

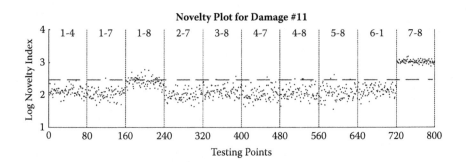

FIGURE 4.24
Novelty index plot for symmetrical approach of damage no. 11.

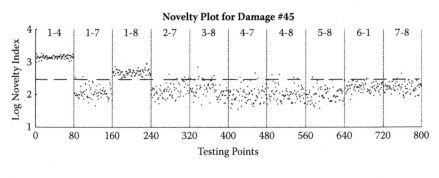

FIGURE 4.25
Novelty index plot for symmetrical approach of damage no. 45.

4.7 Conclusions

This chapter highlights all the results acquired based on the scattering effect of the Lamb wave propagation technique on a CFRP plate under two different structural conditions, namely, the damage and normal conditions. The conducted research work aimed at investigating the feasibility of the acquired waveform response in signaling the true structural conditions of this anisotropic plate (CFRP) under the influence of OA and MLP neural network. The prediction plots generated by adopting the MLP neural network were one of the main results presented in this chapter because the OA alone cannot achieve the desired stage of localizing the potential damage location. Several critical findings can be made based on these OAs or the novelty index plots together with the prediction plots. These findings were thoroughly discussed at the end of each analysis. Only the regression approach together with the time domain and symmetrical approach are considered to be feasible in detecting simulated damage on the CFRP plate by using the force applicator. In summary, the main contention of this chapter serves its purpose by discussing the outcomes of all the experimental works subjected to all of its available generic

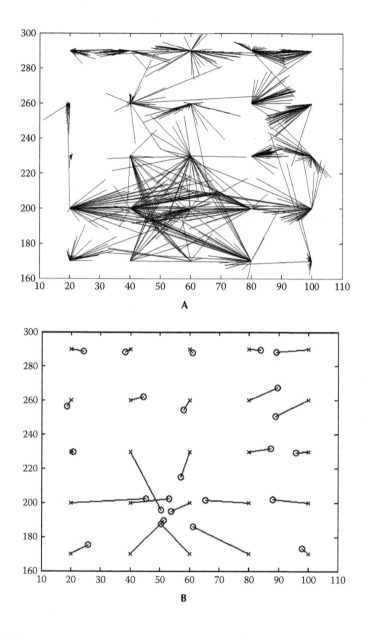

FIGURE 4.26
Hidden layer = 20 with scaling [0 1].

and specific methodologies, strategic procedures, efficient and sufficient data acquisitions, effective feature extraction and condensation, statistical model development, and selected solution algorithms toward fulfilling the damage localization study on composite CFRP plate.

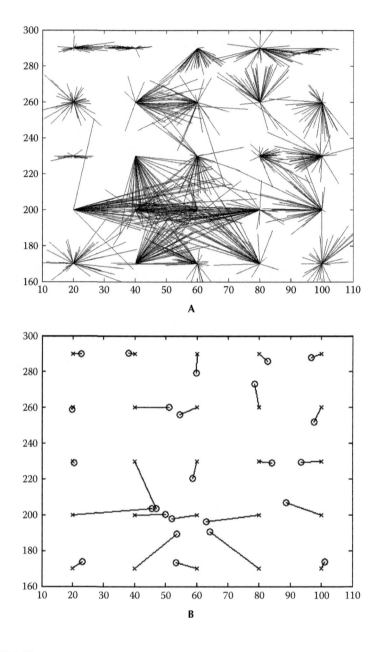

FIGURE 4.27
Hidden layer = 20 with scaling [–1 1].

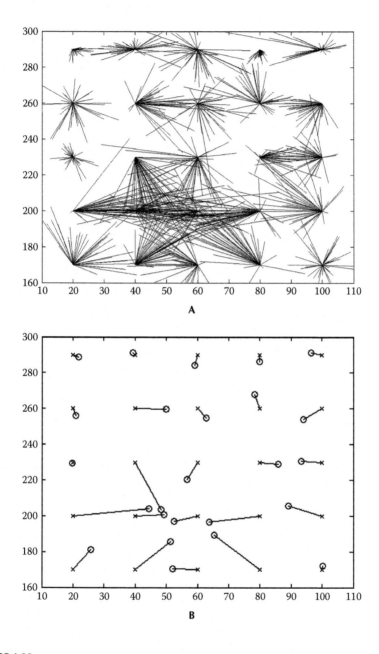

FIGURE 4.28
Hidden layer = 15 and weight decay of 0.1 with scaling [−1 1].

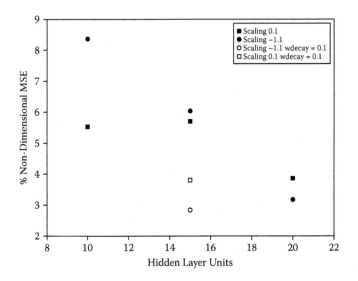

FIGURE 4.29
Nondimensional MSE plot for symmetrical approach.

References

Alleyne, D.N., and Cawley, P. 1992. Optimization of Lamb Waves inspection techniques. *NDT & E International* 25:11 – 22.

Barnett, V., and Lewis, T. 1994. *Outliers in Statistical Data*. Chichster: Wiley.

Bishop, C.M. 1994. Novelty detection and neural network validation. *IEEE Proceeding–Vision and Image Signal Processing* 141:217 – 222.

Bishop, C.M. 1995. *Neural networks for pattern recognition*. New York: Oxford University Press.

Manson, G., Worden, K., and Allman, D. 2003. Experimental validation of a structural health monitoring methodology. Part I Novelty detection on a Gnat aircraft. *Sound and Vibration* 258: 345 – 363.

Markou, M., and Singh, S. 2003. Novelty detection: a review -part 1: Statistical Approaches. *Signal Processing* 83: 2481 – 2497.

Mustapha, F., Manson, G., Pierce, S.G., and Worden, K. 2005. Structural health monitoring of an annular components using statistical approach. *Strain Measurement* 41: 117 – 127.

Nabney, I.T. 2002. *Netlab Algorithms for Pattern Recognition*. London: Springer-Verlag.

Netlab, http:///www.ncrg.aston.ac.uk/netlab (accessed on 07/07/2005).

Rytter, A. 1993. Vibration based inspection of Civil engineering structures., in Building Technology and Structural Engineering. PhD Thesis, University of Aalborg, Denmark.

Sohn, H., Farrar, C.R., Hunter, N.F., and Worden, K. 2001. Structural health monitoring using statistical pattern recognition techniques. *Journal of Dynamics System Measurement and Control* 123:706-711.

Tarassenko, L. 1998. *A Guide to neural computing applications*. London: Arnold.

Tarrasenko, L., Hayton, P., Cerneaz, N., and Brady, M. 1995. Novelty detection for the identification of masses in mammograms. *Proceedings of Fourth IEE International Conference on Artificial Neural Network* 409:442-447.

Worden, K., Ball, A.D., and Tomlinson, G. R. 1993. Fault location in a framework structure using neural networks. *Smart Material Structure* 2: 189 – 200.

Worden, K. 1996. MLP -Multi layer perceptron, *Version 3.4 -User Manuals,* Dynamics Research Group, University of Sheffield, UK.

Worden, K. 1997. Structural fault detection using a novelty measure. *Journal of Sound and Vibration* 201:85-101.

Worden, K., Ball, A.D., and Tomlinson, G. R. 1993. Fault location in a framework structure using neural networks. *Smart Material Structure* 2:189-200.

Worden, K., Manson, G., and Fieller, N.R.J. 2000a. Damage detection using outlier analysis. *Journal of Sound and Vibration* 229: 647 – 667.

Worden, K., Pierce, S. G., Manson, G., Philp, W. R., Staszewski, W. J., and Culshaw, B. 2000b. Detection of defects in composite plates using Lamb waves and novelty detection. *International Journal of Systems Science* 31: 1397 – 1409.

Zapico, J.L., Worden, K., and Molina, F.J. 2001. Vibration-based damage assessment in steel frames using neural networks. *Smart Material Structure* 10:553-559.

5

Damage Localization of Carbon Fiber–Reinforced Plastic Composite and Perspex Plates Using Novelty Indices and the Cross-Validation Set of Multilayer Perceptron Neural Network

F. Mustapha, S. M. Sapuan, K. Worden, and G. Manson

CONTENTS

ABSTRACT   The feasibility of the cross-validation set from the multilayer perceptron (MLP) neural network architecture in localizing damage occurring on both the CFRP composite and Perspex plates is highlighted in this chapter. The partitioned data of the multilayer perceptron neural network that were computed from the novelty indices will be used here to predict the location of the damage occurring on both of the plates. The cross-validation data set introduced here is composed of the training, validation, and the testing data set. It was decided that the training data set will be used in fulfilling the damage localization task. Several rules explained in the methodology section need to be followed to successfully localize the simulated damage occurring on the plates. The novelty values for the overall plate area and the quarter (symmetrical) plate area for both the CFRP and Perspex plates will be tested in this work. The work presented in this chapter is built on the work by Sohn et al. (2001). In their study, a layer of piezoelectric patches is used to generate a wavelet input signal and to capture the response waveform signals. The acquired signals were then passed into the wavelet transform to extract the damage-sensitive features. These features combined with several rules will pinpoint the damage location occurring on the composite plate. As for this work, the tone burst signal will be the input to the system and the defect will be located by using acquired waveform responses computed into novelty indices and combined with several designated rules. At the end of this chapter, a brief comparison between the novelty indices method and the multilayer perceptron neural network will be given, together with a suggestion as to which technique is capable of giving better predictions in terms of localizing the damage on both of the tested plates.

5.1   Introduction to Research Methodology

To identify the damage location, both of the plates were divided into a 12 × 12 mesh grid box as illustrated in Figure 5.1. The size of each cell was 17.9 by 25.4 mm. The actual damage location for damage 1 is represented by the ($x1,y1$) cell as shown below. The dotted rectangular line exhibits the overall damage location region occurring on the plates.

One hundred simulated damage locations were investigated. The first stage of this work involved the overall plate area that was monitored using 28 sensor paths. Therefore, the novelty indices from the 28 sensor paths for all 100 damage locations and 80 observations were kept for further analysis in localizing the damage. Details of the data acquisition processes can be referred to in Mustapha et al. (2005).

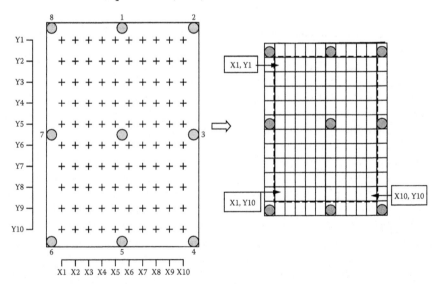

FIGURE 5.1
Damage grids on the plate.

To localize the damage systematically, it was necessary to first divide the novelty indices into a cross-validation set in the ratio of 40:20:20 (training/ validation/testing). The damage location was then predicted from the training data set based on a set of rules. These rules were:

1. Examine the 40 novelty indices (training set) and exclude all the paths that do not exceed the threshold value (i.e., not outlier paths). The novelty index plot is used in this case.
2. Identify outlier paths where 10% of the observations are above the threshold value. All other paths are termed as *inlier paths*. This can be done by outlining the path line and mark the cells that lie along the path.
3. Set the values of all the cells along these paths to 1.
4. Check and add one unit to any cells that contain the intersections of the outlier paths.
5. Highlight the cells that only score more than one unit. (All the cells that have more than one intersection between the outlier paths will be highlighted.) If only one cell contains the highest score or only one cell has a score more than one unit, select this respective cell to be the predicted damage location.
6. Check for inlier paths for these highlighted cells.
7. A cell that does not lie on any inlier paths and contains the highest score will be selected as the damage point.

We also need to take note of step 5; if only one cell exists with a score more than one unit, the damage localization process ends here and the respective cell is likely to be the predicted damage location. Several damage points were selected and tested to demonstrate the effectiveness of these rules, and the results will be presented in the next section.

Figure 5.3a and b demonstrate the capability of the technique in localizing the damage on the designated position (in this case, damage 15). Figure 5.3a and b adopt the first two rules set earlier, where only the first 40 training data samples were used and the novelty index plot of Figure 5.2 is then used to determine which paths are outliers. Paths such as 1-5, 1-6, and 1-7 are automatically included as outlier paths and path 1-3 is included because of the rules of step 2. By implementing this step, all the selected paths were outlined and the cells lying on these paths are then marked by gray areas, as shown in Figure 5.3b.

Figure 5.4a and b indicate the cells condition after following the set of rules in steps 3–4. Step 4 allows the checking and adding of one unit to any cells that contain a path intersection. As a result, the cell that had the most intersections scored 3 units (see Figure 5.4a). It should be noted that only cells with scores of 2 or 3 will be the focus here to proceed further with step 5.

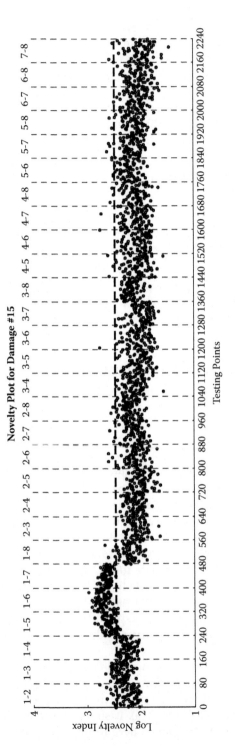

FIGURE 5.2
Novelty index plot for the respective damage location.

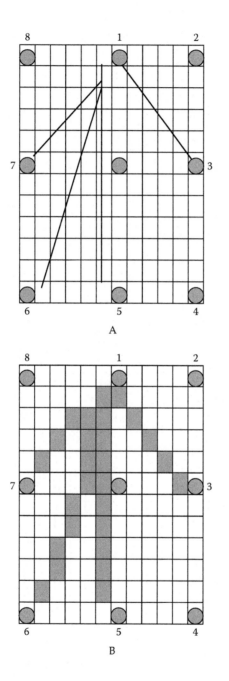

FIGURE 5.3
Results of undertaking steps 1–2.

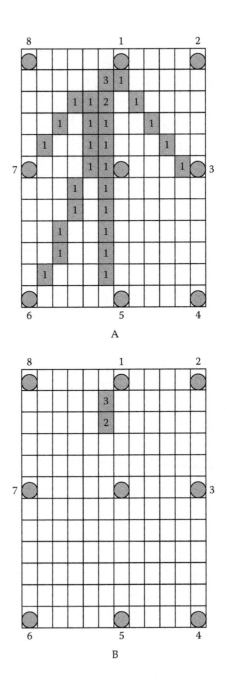

FIGURE 5.4
Results of undertaking steps 3–5.

Because there are two probable points of damage location, one needs to follow step 6 as this vital step will get rid of or cancel out any cells that lie on the inlier path (i.e., path 1-8; please refer to the novelty index plot in Figure 5.2). Because of this step, the highest score cell is dismissed as being the potential damage point, leaving one cell with a score of 2, as shown in Figure 5.5b. Based on this technique, there was only one point indicated as a probable damage location. All of the training data samples satisfied all the sets of conditions in the rules and pointed to the actual damage location (Figure 5.6). Further analysis was extended to damage location of $(x3,y3)$ (see Figure 5.7) to examine how the method performed with an alternative damage location. It is essential to test this damage localization technique on a variety of different damage locations. The novelty index plot for this new damage location is shown in Figure 5.8. By using similar procedures and steps as those conducted previously, the outcome of this damage localization process is then shown in Figures 5.9a to 5.10b.

For this attempt, the method suggests only one potential damage location, and this corresponds exactly with the true damage location shown in Figure 5.11. As highlighted in the methodology section, the proposed method recommends that if only one cell has the highest score after steps 1 to 5 have been undertaken, then that cell should be selected as the predicted damage location. This is why the damage location in Figure 5.11 was accurately predicted. Based on these two results for the prediction of damage at locations 15 and 23, the set of rules earlier is therefore effectively able to predict the exact damage point on the isotropic plate. To see whether the method can perform on both the overall plate area and the quarter plate area, the following section will present the findings for damage location 15 when only the quarter plate area was used.

5.2   Damage Localization on the Quarter Plate Area

Similar rules and mesh grids were applied in this section, with the exception of less coverage area and the inclusion of the middle sensor. Only 25 damage locations were used together with 10 sensor paths (see Figure 5.18) storing the novelty index values. The true damage location area for this case is within the dotted rectangular line that can be seen in Figure 5.12.

For comparison purposes with the previous section, only one damage location was again used here, this being location 15 (see Figure 5.13).

The approach of lesser coverage area was introduced here with the inclusion of the middle sensor, which was tested for damage localization using the novelty indices method. As for the previous case using the overall plate area, this method again successfully predicted the single damage at location 15 (see Figure 5.14).

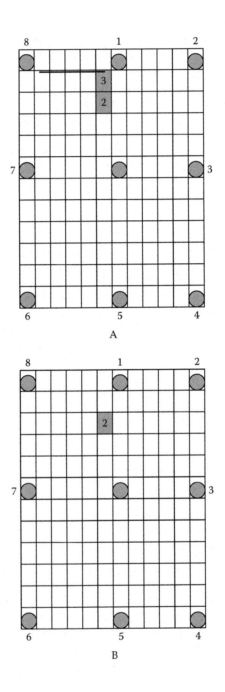

FIGURE 5.5
Using steps 6–7 to find the probable damage location.

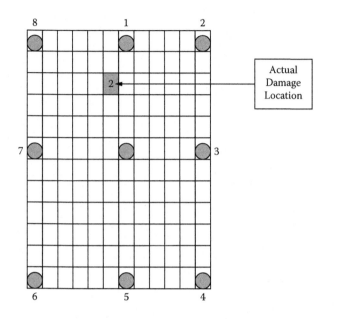

FIGURE 5.6
Damage location occurring on the isotropic plate.

To see whether or not this technique can be applied to an anisotropic plate, damage 23 will be used here; and because the symmetrical approach gives better resolution on the novelty index values, only the symmetrical approach with damage 23 will be tested for the anisotropic plate structure (Figures 5.15 and 5.16).

5.3 Damage Localization on the CFRP Plate

For this case, the technique begins with an observation of the novelty index plot for the specified damage location. Because the symmetrical approach generates reasonable results for the OA and multilayer perceptron (MLP) results, the damage localization study will limit only to composite plate using only quarter plate area. The novelty index plot can be seen in (for the case of a symmetrical approach containing damage at location 23).

Figure 5.17 shows that the method incorrectly predicts the damage location occurring on the anisotropic plate. The method identifies two path intersections that are both undamaged and then incorrectly selects a single undamaged cell. This is because some strange paths go above the threshold for the case of anisotropic plate used here.

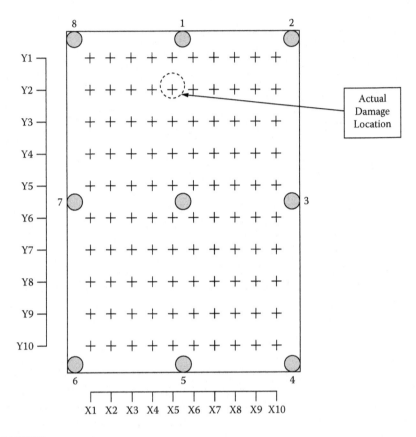

FIGURE 5.7
Actual damage location of $(x3,y3)$.

5.4 Conclusions

In this chapter, an alternative means of localizing the damage occurring on both of the plates has been introduced. The method only utilizes the novelty index values from the network of sensor paths on the plates, together with the 40 novelty indices from the training data set. Based on the plotted results, this damage localization technique can be described as reliable for the isotropic plate but not for the CFRP plate. Thus, the technique of combining the novelty indices and the MLP neural network that was adopted in previous chapter appears to be more effective for localizing damage in both isotropic and anisotropic plates than the technique introduced in this chapter. The difference in performance is shown in Figures 5.18 and 5.19. For case 1, an MLP neural network mapped the prediction value of damage 15 on an x- and y-coordinate location without relying on any rigid set of rules. The solution predicts on a qualitative value rather than directing into a single cell

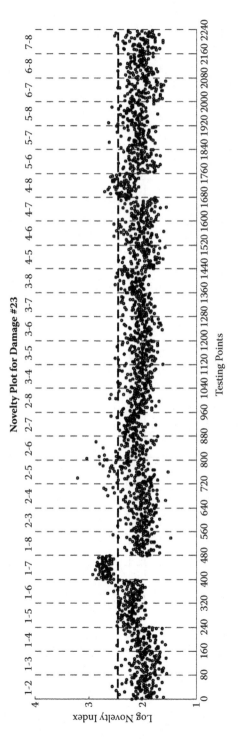

FIGURE 5.8

Novelty index plot for the (x3,y3) corresponds to damage no. 23.

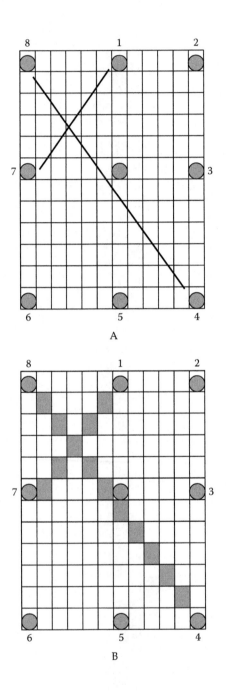

FIGURE 5.9
Results of undergoing steps 1–2.

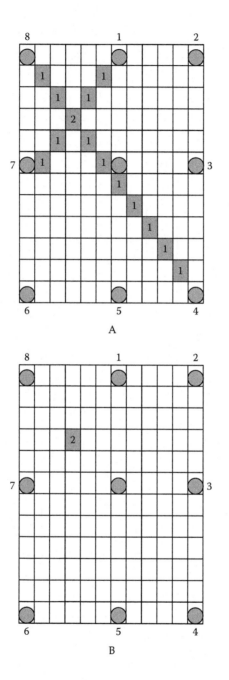

FIGURE 5.10
Results of undergoing steps 3–5.

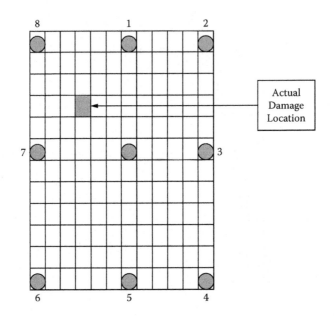

FIGURE 5.11
Damage location on the isotropic plate for $(x3, y3)$.

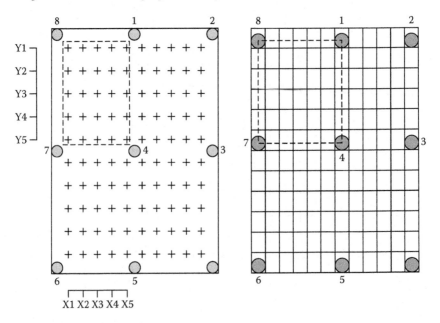

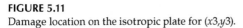

FIGURE 5.12
Damage grids on the plate for the symmetrical approach.

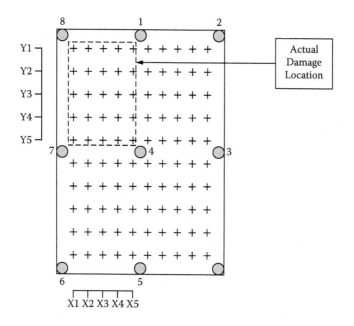

FIGURE 5.13
Actual damage location 15 (x5,y2) using the symmetrical approach on the isotropic plate.

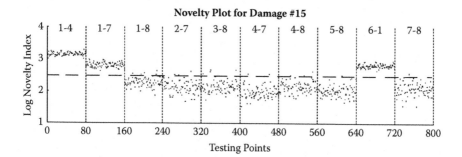

FIGURE 5.14
Novelty index plot for the (x5,y2) damage location corresponding to damage no. 15 and using a symmetrical approach for the isotropic plate.

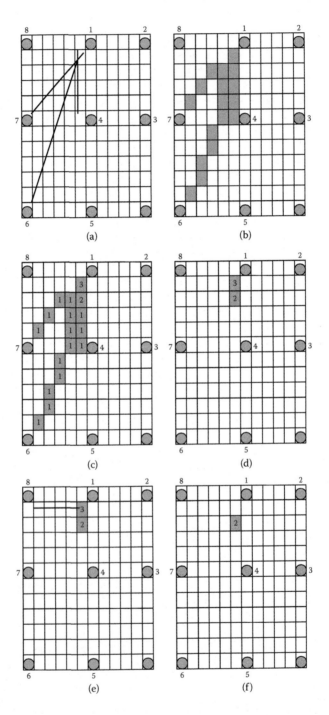

FIGURE 5.15
Results of damage localization using the proposed method with damage at location 15.

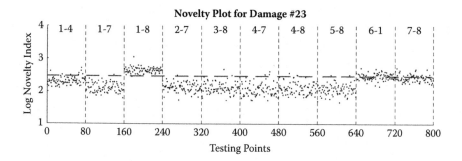

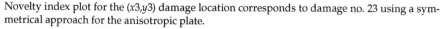

FIGURE 5.16
Novelty index plot for the $(x3, y3)$ damage location corresponds to damage no. 23 using a symmetrical approach for the anisotropic plate.

compound. The MLP approach would undoubtedly facilitate the process of localizing the individual damage occurring on the plate.

As the MLP neural network technique is a regression approach to damage localization, it is able to give a more accurate prediction of the damage location than the novelty index when with a cross-validation set (see Figure 5.19 for the isotropic plate only under the symmetrical approach). Based on these findings, the MLP neural network approach is a more effective candidate for realistic and systematic damage localization on both types of plate structure, namely, the Perspex and the CFRP.

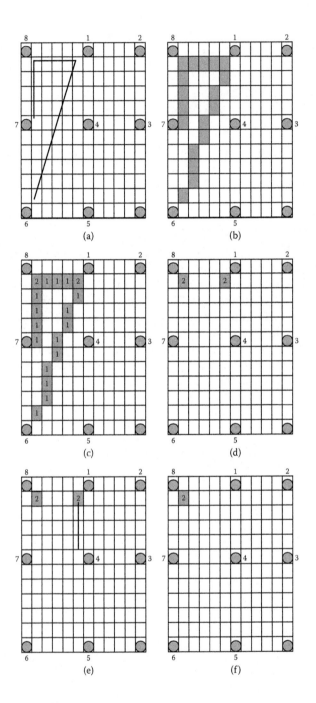

FIGURE 5.17
Results of damage localization using the proposed method on damage no. 23 for anisotropic plate.

<table>
<tr><th>Plate Properties</th><th>MLP Neural Network</th><th>Novelty Indices with Cross Validation Set</th></tr>
<tr><td>Isotropic</td><td>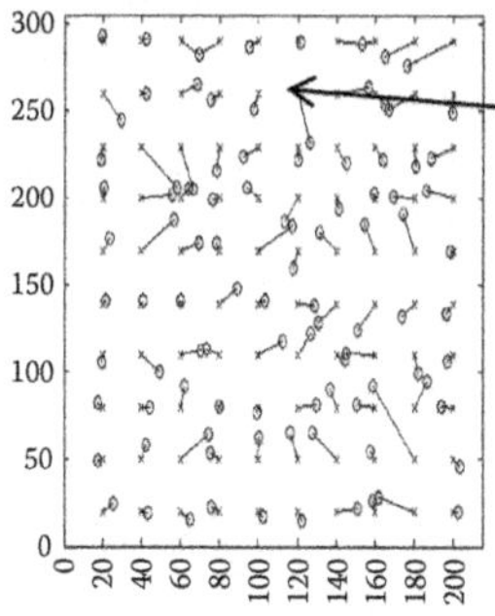
</td><td>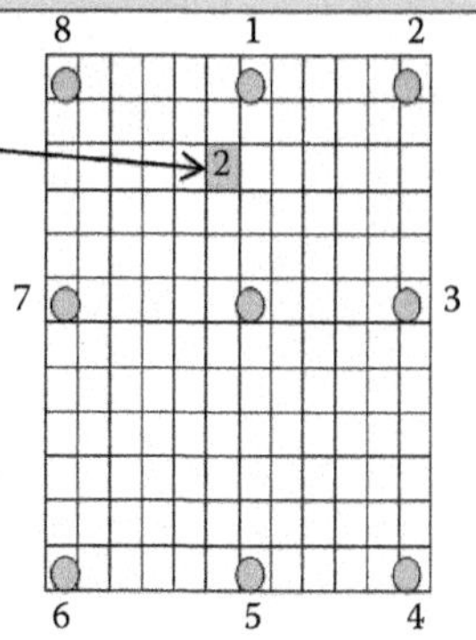
</td></tr>
</table>

FIGURE 5.18
Results of overall plate area on both of the techniques (case 1: overall plate area).

<table>
<tr><th>Plate Properties</th><th>MLP Neural Network</th><th>Novelty Indices with Cross Validation Set</th></tr>
<tr><td>Isotropic</td><td>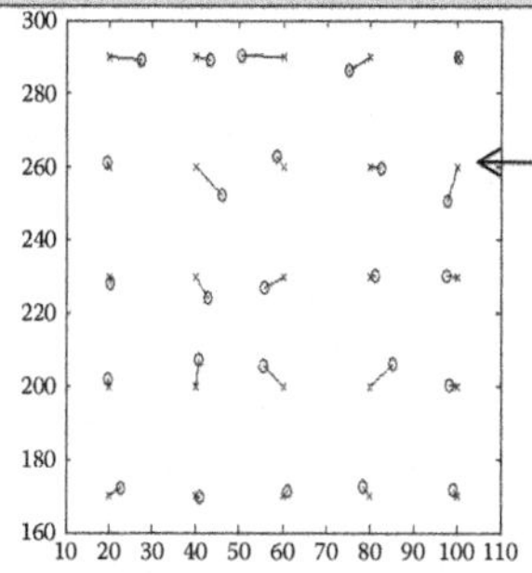
</td><td>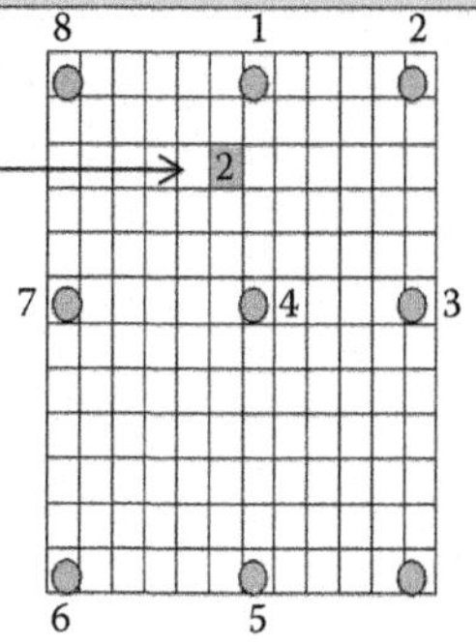
</td></tr>
</table>

FIGURE 5.19
Results of both of the techniques on the quarter plate area (case 2: quarter plate area (symmetrical approach).

References

Mustapha, F., Manson, G., Pierce, S. G., and Worden, K. 2005. Structural health monitoring of an annular component using a statistical approach. *International Journal of Strain Measurement* 41:117–127.

Sohn, H., Farrar, C. R., Hunter, N. F., and Worden, K. 2001. Health monitoring using statistical pattern recognition techniques. *Journal of Dynamics System Measurement and Control* 123:706–711.

6

Impact Damage Detection in a Composite Structure Using Artificial Neural Network

S. Mahzan and W. J. Staszewski

CONTENTS

ABSTRACT Composite materials have been extensively used in many engineering applications due to their high specific strength and modulus, together with their flexibility in design. The susceptibility of composite materials to incur impact damage creates a major concern related to structural integrity, especially hidden damage caused by low-velocity impacts and fatigue. Sensitive and reliable damage detection methods are needed to prevent any damage-related problems. This chapter briefly discusses a method of impact damage localization in a composite structure using an advanced signal processing procedure associated with an artificial neural network. This method uses the regression-based impact localization using the multilayer perceptron neural network. The specimen used for the study was a composite wing box structure and utilized nine piezoceramic transducers. Impacts were introduced on the composite structures at predefined points. Several impact features from impact's strain data were used for neural network training, validation, and testing. Results from neural network processing demonstrate the comparative values between the actual and estimated coordinates. The percentage of error was found to be less than 5%. This shows that the artificial neural network is capable of estimating impact damage location in a composite structure.

6.1 Introduction

Composite materials have been widely used in many engineering applications covering various areas of land, air, and sea transportation. Their high specific strength and modulus, together with their flexibility in design, have made composites a very good choice for structural design, compared to traditional materials such as metals. Composites, with their load carrying capabilities, low density, and superior performance, are particularly needed for the latest emerging structures in aerospace.

The good performance displayed by composites has benefited many industries, especially the transportation sector. Composite materials, apart from their strength and low weight, also offer resistance to fatigue, corrosion, and impact damage. This is particularly important in aerospace engineering. Therefore, many aircraft manufacturers, such as Boeing and Airbus, have implemented composite materials in their products. For example, in a Boeing B-777, several structural components are manufactured from composite materials such as graphite, fiberglass, and hybrid composites, comprising about 12% of the total materials used in the structure (Walz, 2006). Application of composite materials in the newly designed Boeing B-787 is also increased to almost 50% of the total materials used (Walz, 2006). The examples show that composite materials are popular and have become more important in advanced engineering structures. It is inevitable that the future will lead to more advanced structures with new types of complex composite materials. Understanding the properties and behavior of these materials, together with possible failure mechanisms, will be of great importance to designers and end users of structures.

The susceptibility of composite materials to incur impact damage is well known and is a major concern related to structural integrity. Low-velocity impacts are often caused by bird strikes, tool drops during manufacturing and servicing, or runway stones during takeoff. Such impacts may result in various forms of damage such as indentation, delamination, or fiber/matrix cracking, leading to severe reduction in strength and integrity of composite structures. Although structures designed with safe-life principles can withstand in theory catastrophic failures, impact damage detection is an important problem in maintenance of aircraft and space structures (Mal et al., 2005). Visible damage can be clearly detected and remedial action taken to maintain structural integrity. However, a major concern to end users is the growth of undetected, hidden damage caused by low-velocity impacts and fatigue. This undetected, hidden damage is also known in aerospace applications as barely visible impact damage (Staszewski et al., 2002). Failure to detect barely visible impact damages may result in a catastrophe. Therefore, sensitive and reliable damage detection methods in composite materials are needed to prevent damage-related problems.

This chapter demonstrates an application of advanced signal processing procedure for impact damage detection in a composite structure. The advanced signal processing used in this work was based on an artificial neural network (ANN). The main specimen used was an aircraft composite wing box structure with a dimension of 960 × 660 mm.

6.2 Impact Damage Detection in Composite Structure

Damage detection and localization play an important role in the entire damage identification process at the earliest stage of monitoring for damage. Several different approaches can be implemented for impact damage detection, including techniques based on nondestructive testing/evaluation and structural health monitoring technologies. These techniques use passive or active approaches. The active approach is generally based on actuating and sensing of structures and includes such techniques as ultrasonics, acousto-ultrasonics, Lamb wave inspection, x-ray, and shearography (Kessler et al., 2002; Diamanti et al., 2004; Prasad et al., 2004; Toyama and Takatsubo, 2004; Grondel et al., 2004; Wang and Chang, 1999; Sekhar et al., 2006). Various signals are used to interrogate the structure within these techniques. Signal arrival times for the respective damaged and undamaged conditions were calculated from scatter waves and compared. These comparative measurements were then used to investigate for any possible damage, that is, location and size, present on the composite structure.

Unlike active methods, passive techniques do not involve actuator transducers. A series of receiver sensors are used to "sense" and/or "hear" any perturbation caused by damage. Various approaches are possible. For example, in impact damage detection, sensors are either embedded into or bonded onto structures in order to monitor (i.e., detect and locate) impact strain data. The energy of impacts is then estimated using advanced signal processing techniques. It is assumed that certain energy levels can lead to structural damage. This information is available from material and design studies. Acoustic emission, used in many engineering applications, is a passive method. Several examples of acoustic emission used in passive damage detection can be found in the work of Yang and Han (2002), Wolfinger et al. (1998), Dupont et al. (1999), and Kirikera et al. (2007).

Impact damage detection can be used to estimate impact location and energy. The estimation of impact location and energy/force can be obtained by using advanced signal processing techniques, namely, the ANNs. Worden and Staszewski (2000) applied an ANN to estimate impact location in a rectangular composite plate. They used four data features: time of the maximum response after impact, magnitude of the maximum response, peak-to-trough range of the response, and real and imaginary parts of the response spectrum.

The results showed that an ANN was able to locate the impact damage with an error area of 1.5%. Staszewski et al. (2000) used similar data features for impact location in a composite structure. Here, a noise level of 20% of the measured strain data was introduced to training data. Impact locations were identified with a percentage error per analyzed area of 2.5%.

Sung et al. (2000) implemented an ANN for impact location on composite and aluminum plates. The network used for the research was a multilayer perceptron (MLP) with a fast learning algorithm (Levenberg-Marquardt) and generalization methods (i.e., regularization and early stopping). The arrival time of acoustic emission signals was measured from the leading wave of the signal. The differences between arrival times from four sensors were used as the network input. The results showed that impact locations could be detected with an average error of less than 5 mm on a 330 × 330 mm composite plate.

Haywood et al. (2001, 2005) investigated the potential of SMART Layer® sensors and an ANN-based algorithm to locate impacts. Six different data features from the original strain signals, spectrum, and envelopes were used. It was found that impact localization using an ANN was better predicted using the arrival time and maximum value of the analyzed envelope functions calculated from the strain data. The average impact location error of this investigation was about 0.85% of the total area of the plate. Le Clerc et al. (2004, 2007) used similar data features to locate impacts in a more complex structure, that is, a composite wing box structure from a commercial aircraft. Results showed that an ANN was capable of locating impact damage in a complex structure. The average error in this study was 1.92%. This result is particularly good considering the geometrical complexity of the structure.

Recently, work by Lee et al. (2006) proposed ANN-based damage detection using long continuous sensors. The method utilized strain signals from impact or damage acquired by long sensors. It was observed that long continuous sensors do not miss any damage responses. The network used the maximum amplitudes of strain signals as the input, and the network outputs were the damage locations. The network was capable of effectively locating damage; however, this method just focused on impact localization. The impact/damage size was not estimated.

6.3   Impact Location with Neural Networks

Various network architectures can be used in practice. In this work, a mapping procedure based on ANNs is used for impact location in composite structures. More specifically, the MLP network is chosen. Following Worden (1996), a brief introduction to the MLP network is given in this section.

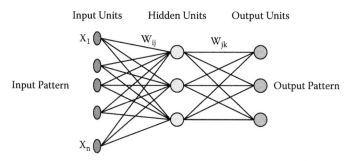

FIGURE 6.1
General architecture of a multilayer perceptron (Tarassenko, 1998). With permission.

The MLP network is the most widely used network architecture in engineering applications. The MLP works in a multilayer, feed-forward structure, and consists of a series of connected nodes arranged in layers. These are the input layer, hidden layer, and output layer nodes. The MLP network is trained by a supervised learning scheme, and therefore requires desired responses in order to be trained. A general MLP architecture is illustrated in Figure 6.1.

The MLP process starts as the input patterns enter at the input layers $(x_1, x_2, ..., x_n)$ units. These input patterns progress forward through the hidden layers and the results emerge at the output layer as an output pattern. Each node i in the preceding layer is connected to each node j at the following layers through a connection weight w_{ij}. This weight is used to scale the output signal from the node i as it enters the node j. For each node i in layer k (a hidden layer), a weighted sum for all signals $x_j^{(k-1)}$ from the preceding layer $(k - 1)$ is calculated, giving the excitation $z_j^{(k)}$ of the node. This excitation is then passed through a nonlinear activation function f to emerge as the output value $x_j^{(k)}$ to the next layer, as given by:

$$x_i^{(k)} = f\left(z_i^{(k)}\right) = f\left(\sum_j w_{ij}^{(k)} x_j^{(k-1)}\right)$$

(6.1)

The hyperbolic tangent (tanh) is often used for the activation function f, that is, $f(x) = \tanh(x)$. The output of this function is in the range $(-1, 1)$, as illustrated in Figure 6.2.

Each network layer has a bias node that is connected to all other nodes in that layer. The output of each bias node is held fixed in to allow for constant offsets in the excitations z_i of each node. The x_i is then transformed by the second layer of weights and biases to give:

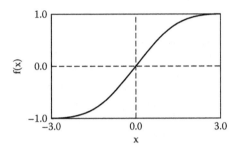

FIGURE 6.2
The hyperbolic tangent activation function, $f(x) = \tanh(x)$.

$$x_i^{(k)} = f\left(z_i^{(k)}\right) = f\left(\sum_j w_{ij}^{(k)} x_j^{(k-1)} + b_j^{(k-1)}\right) \tag{6.2}$$

Finally, the above values are then passed throughout the output unit activation function to give the output values y_k.

6.4 Methodology

Regression-based impact localization problems compare predicted outputs with actual inputs using continuous values. Here, the network inputs are the actual locations of impact data features, given by x and y coordinates, and the network outputs are given by estimated x and y coordinates for the impact location.

Impact events on a composite structure produce strain waves that can be captured by a network of transducers. Several data features can be obtained from these impact strain waves, for example, maximum peak, minimum peak, and peak-to-peak values, and the MLP networks use these data features as inputs. Since MLP networks are obtained through three stages, that is, training, validation, and testing, three different sets of data features are therefore required for impact location analysis. The purpose of using these different data sets is to ensure that the network is capable of generalization.

The MLP network starts with the training stage. At this stage, the network's architecture is examined to determine the optimum number of hidden units, with corresponding sets of connection weights w_{ij} for each arrangement. This stage utilizes data features from the training set and corresponding target output values (actual impact positions x and y for regression). The connection weights are iteratively adjusted until the network produces the best match for the actual impact positions. The results of this process depend on the initial values given to the weights. Therefore, several training sessions with different initial weights are required for each given network architecture. The impact location performance was assessed using a percentage error expressed in terms of the total area of the plate, given by

$$\text{Nondimensional MSE, } \varepsilon_A = \frac{\sum_{i=1}^{N} |\hat{x}_i - x_i| \times |\hat{y}_i - y_i|}{\text{area of structure} \times (N)^2} \times 100\% \qquad (6.3)$$

where, for a given total of N impacts on the composite structure, Art was here, but was deleted. and Art was here, but was deleted. are the predicted coordinates, x and y are the actual coordinates, and x_p and y_p are the size structure in the x and y coordinate axes.

Once the training stage is completed, the next step is to select the best network architecture using the validation set. Once the optimal network is constructed, the test set is used to estimate the impact locations. The difference between the actual and estimated impact coordinates was calculated as an average distance. The percentage error of average distance per axis can be calculated as

$$E_s(x) = \frac{\sum_{i-1}^{N} |\hat{x}_i - x_i|}{(N) \times x_p} \times 100\% \qquad (6.4)$$

and

$$E_s(y) = \frac{\sum_{i-1}^{N} |\hat{y}_i - y_i|}{(N) \times y_p} \times 100\% \qquad (6.5)$$

The entire procedure for impact location in a composite structure is shown as a flowchart in Figure 6.3.

6.5 Test Specimen—an Aircraft Composite Wing Box Structure

The test specimen used in the experimental impact localization work was a composite wing box structure, shown in Figure 6.4. The structure was a

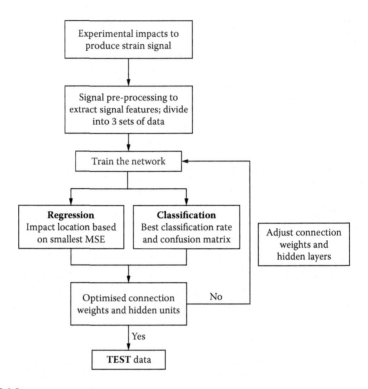

FIGURE 6.3
MLP network procedures for impact location in a composite structure.

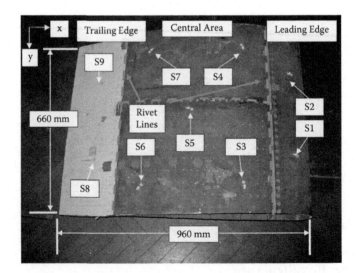

FIGURE 6.4
Composite aircraft wing box structure used for impact damage location.

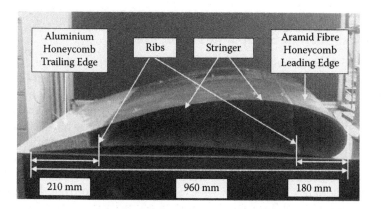

FIGURE 6.5
Cross-sectional view of the composite aircraft wing box structure used for impact damage location.

section of one of the flaps taken from a commercial aircraft. The width and length of the structure were approximately equal to 960 and 660 mm, respectively. The structure had a complex geometry and was manufactured from different materials. Figure 6.5 shows the cross section of the composite structure. The composite skin of the wing box was curved at the top. The leading and trailing edges were manufactured as sandwich structures with aramid fiber and aluminum honeycomb, respectively. The central area had numerous interior ribs and stringers. One horizontal line and two vertical lines of rivets were present on the top curved surface of the wing box. These lines also indicated positions where underside ribs and spars were attached.

The structure was instrumented with nine Sonox-P5 piezoceramic sensors (diameter 6.5 mm, thickness 0.25 mm) bonded onto the top surface. Two sensors were located on the trailing edge, five on the central area, and two on the leading edge. The sensor positions, labeled as S1, S2, …, S9 in Figure 6.4, are given in Table 6.1.

TABLE 6.1

Sensor Locations on the Composite Wing Box Structure

Sensors	x Coordinate (mm)	y Coordinate (mm)
S1	868	470
S2	868	150
S3	667	577
S4	670	58
S5	485	325
S6	317	588
S7	305	70
S8	115	494
S9	105	182

6.6   Experimental Setup and Procedure

A series of impacts were performed on the aircraft composite wing box structure. The impacts were performed using a modal impact hammer, in order to guarantee low-velocity, low-energy impacts that did not damage the structure. The impact strain data were sensed using nine piezoceramic transducers. The strain data were collected using an nCode nSoft v. 5.2 acquisition system controlled by an nCode dAtagate v. 5.1 software and running on a Dell Inspiron 3500 laptop PC. The data acquisition was triggered when strain signals received by the sensors exceeded a threshold level of 0.1 V. A sampling frequency of 6250 Hz was used for analog-to-digital conversion and an overall time of 2 s was used for data acquisition.

Strain signals were collected for impacts performed both on a regular grid and at random positions on the structure. Figure 6.6 shows sensor and impact positions on the aircraft composite wing box structure, where the impact locations were given by x and y coordinates. All this information is summarized in Table 6.2. Altogether, 1733 strain wave signals were captured and stored on the PC's hard disk for further analysis. The strain data were initially low-pass filtered using a sixth-order Butterworth filter, implemented in MATLAB v. 6.5, in order to remove undesired noise. The normalized cutoff frequency was equal to 0.1. The mean was then removed from all strain signals.

Figure 6.7 gives examples of strain data acquired by eight different sensors for an impact event performed at $x = 365$ mm, $y = 505$ mm (position A in Figure 6.6a). This impact event occurred in zone 2, very close to sensor S6; hence, the maximum peak-to-peak amplitude was exhibited by the strain data from sensor S6 in Figure 6.7. The following maximum peak-to-peak amplitude was measured by sensor S3, which was located in the same zone as sensor S6. Generally, the results illustrate that the farther the sensor was positioned from the impact location, the smaller the amplitude of the strain wave, as expected. The strain data recorded in this section are used for an ANN-based impact location procedure in a later section.

6.7   Network Design

The first implementation step was the design of the MLP network. The network was designed and trained following the regression algorithm, where the back-propagation algorithm was used for training. The input to the network consisted of a mixture of the seven signal features selected from the strain data, that is, the maximum peak, minimum peak, peak-to-peak, and variance in the time domain, and the arithmetic mean from the absolute, imaginary, and real spectra in the frequency domain.

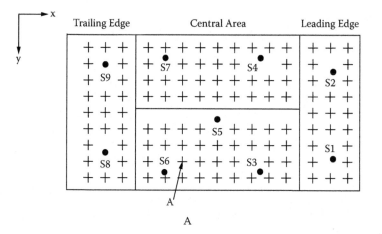

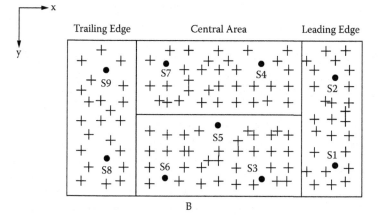

FIGURE 6.6
Impact positions for the aircraft composite wing box structure on a: (A) regular grid, (B) random grid. Sensor positions are indicated by the black-filled circles.

Three different MLP architectures were implemented in this study. The first of these architectures used a single feature as the input. Seven different networks were initially designed for the seven signal features named above. For each different signal feature, strain data from nine sensors were used. For the second network architecture, the combination of all time-domain features used 36 elements (four strain data features from nine sensors), whereas for the combination of all frequency-domain features, the networks consisted of 27 elements (three strain data features from nine sensors). Finally, the third network architecture used all seven time- and frequency-domain features together. This network was fed with an input of 63 elements (seven strain

TABLE 6.2

Summary of Impact Events Applied to the Analyzed Zones of the
Aircraft Composite Wing Box Structure

Zone	Regular Grid Size (mm)	Total Number of Impacts	Impacts (Regular Grid)	Impacts (Random)
1	30 × 60	425	255	170
2	25 × 40	420	280	140
3	30 × 40	420	252	168
4	25 × 25	468	312	156

data features from nine sensors). The output of the network always gave two elements: the estimated x and y coordinates of the impact.

The overall impact data collected in the experimental tests were divided into three different sets. The training data set used data features from random impacts, whereas the remaining impacts from the regular grid were divided into a validation set and a testing set. The impact location analysis using the MLP involved three different steps. First, the network was trained using the training data set as the inputs, and the connection weights were iteratively adjusted until the network produced the best match to the actual impact positions.

The second step validated the network using the validation data set. At this stage, the most suitable number of hidden layers and hidden layer nodes was determined for each network. Since there is very little guidance regarding the optimum number for these quantities, the initial test was repeatedly performed with one hidden layer of 20, 25, and 30 hidden nodes, respectively. Previous studies demonstrate that one layer is sufficient to model any regression problem (Haywood et al., 2001, 2005). The network was trained to estimate impact coordinates for a single signal feature as the input, and the MSE, defined by Equation (6.3), was calculated for these features. The MSE results are given in Figure 6.8. This clearly shows that an MLP with 25 hidden layer nodes produced the smallest MSE for zones 1, 2, and 4. Therefore, one hidden layer with 25 elements was used for impact estimation. However, for zone 3, an MLP with 20 hidden layer nodes produced the smallest MSE. Hence, this quantity was chosen as the optimum hidden node number for impact location.

Signal features selected from the strain data—that is, the maximum peak, minimum peak, peak-to-peak, and variance in the time domain, and the arithmetic mean from the absolute, imaginary, and real spectra in the frequency domain—were used as input for impact estimation in a wing box structure. Finally, once the optimal network had been constructed and validated, the testing data set was used to estimate impact locations. This led to the final estimated results, which were compared with the actual impact coordinates.

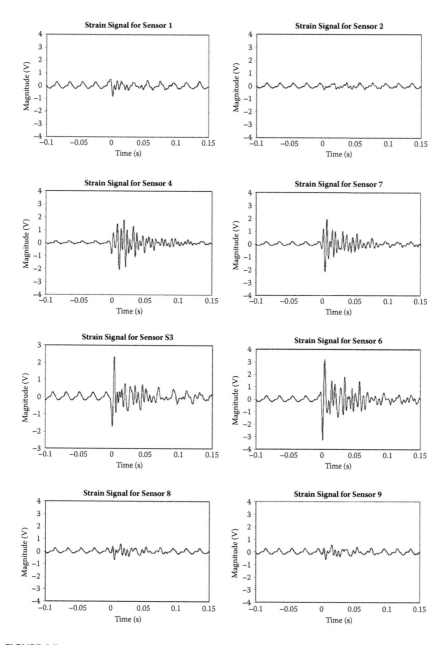

FIGURE 6.7
Examples of strain wave signals acquired by various sensors for an impact performed at $x =$ 365 mm, $y = 505$ mm (position A in Figure 5.6a).

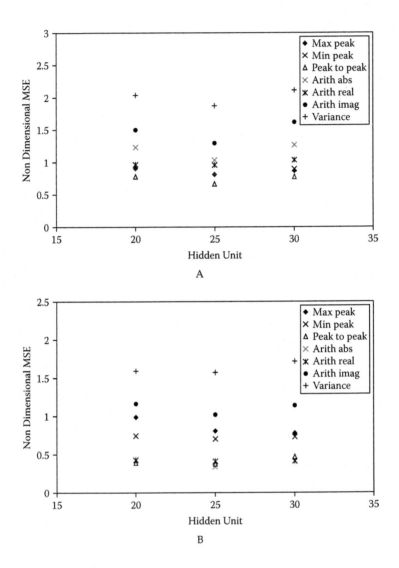

FIGURE 6.8
Hidden unit selection for the aircraft composite wing box structure analyses for: (A) zone 1, (B) zone 2, (C) zone 3, and (D) zone 4.

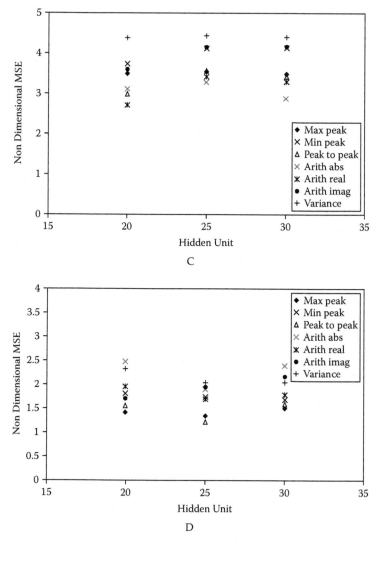

FIGURE 6.8 (CONTINUED)

6.8 Impact Location Results—Local Impact Detection Analysis

Local impact detection refers to detection of impact damage within a small area, predefined as zone 1, zone 2, zone 3, and zone 4. The MLP analysis involved three different network architectures: single feature as the input, combination of all time-domain features and/or combination of all frequency-domain features, and finally all seven time- and frequency-domain features together. Figure 6.9 demonstrates the impact location results of zone 1 when the individual input vector of the maximum peak (time domain) was used. The comparison of actual and estimated impact coordinates for zone 2 using the individual absolute spectrum arithmetic mean feature (frequency domain) is shown in Figure 6.10.

Figures 6.11 and 6.12 show the impact location results for zone 4 and zone 1 when using the combination of all time-domain and all frequency-domain features, respectively. The results show that the overall trend for the estimated coordinates is maintained if compared with the relevant actual coordinates for all parameters investigated.

Finally, the third network architecture applied was the combination of all time- and frequency-domain features together. The impact location results are shown in Figure 6.13. The study demonstrates that ANNs produced good impact location estimates. Table 6.3 summarizes the impact estimation errors for all signal features used for a wing box structure. From the table, when the ANN-based procedure was used, the results are improved significantly only for zone 3 when all features are used for training. The maximum error area was reduced from 4.3% to 2.39%. The ANN-based procedure has produced the best results from the central area (zones 1 and 2) of the wing box structure. It was observed that impact estimation errors were less than 0.7% when peak-to-peak values or all features were used for training. This result was expected since zones 1 and 2 were not filled with honeycombs, which cause significant wave attenuation. The results for zone 3 are unexpected and need further investigation. The results for zone 2 were expected to be better than for zone 3, as explained above.

6.9   Impact Location Result—Global Impact Detection Analysis

The previous impact location analysis has concentrated on local impact detection, that is, the MLP has been trained with the strain data from a separate, predefined structural zone. However, in contrast to the previous local analysis, this impact detection was performed globally in an aircraft composite wing box structure. The term "global" means that the impact detection is conducted using the strain data from the entire structure for network training.

Similar signal features from the strain data, in time and frequency domains, were used for network design. Similarly, the three network architectures were implemented for global impact detection. The overall impact data collected in the experimental tests were divided into three different sets. The training data set used data features from random impacts (526 points), whereas the remaining impacts from the regular grid were divided into a validation

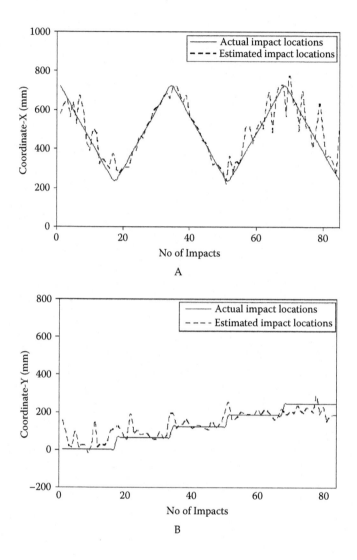

FIGURE 6.9
Comparison between the actual and estimated impact coordinates for zone 1 when maximum peak signal feature is used: (A) *x* coordinate, (B) *y* coordinate.

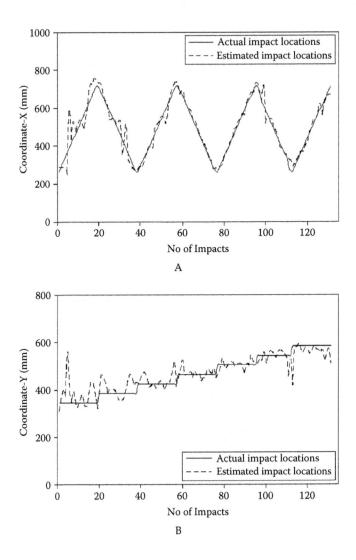

FIGURE 6.10
Comparison between the actual and estimated impact coordinates for zone 2 when the absolute spectrum arithmetic mean feature is used: (A) *x* coordinate, (B) *y* coordinate.

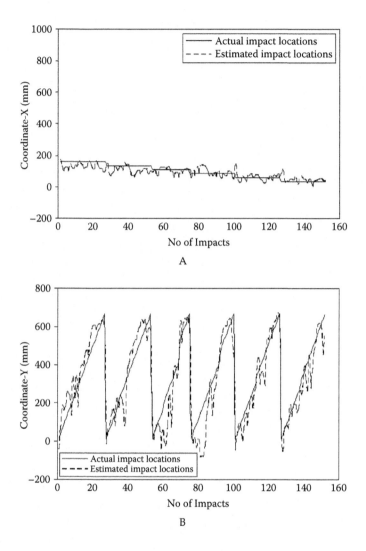

FIGURE 6.11

Comparison between the actual and estimated impact coordinates for zone 4 when all four time-domain signal features are used: (A) x coordinate, (B) y coordinate.

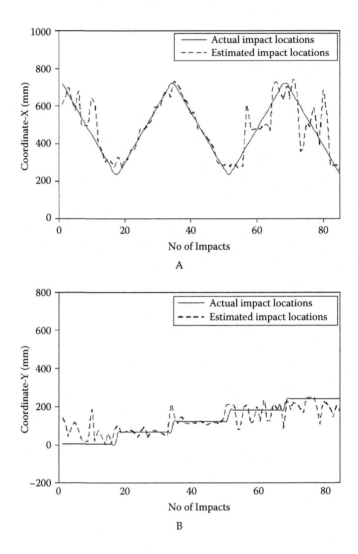

FIGURE 6.12
Comparison between the actual and estimated impact coordinates for zone 1 when all three frequency-domain features are used: (A) *x* coordinate, (B) *y* coordinate.

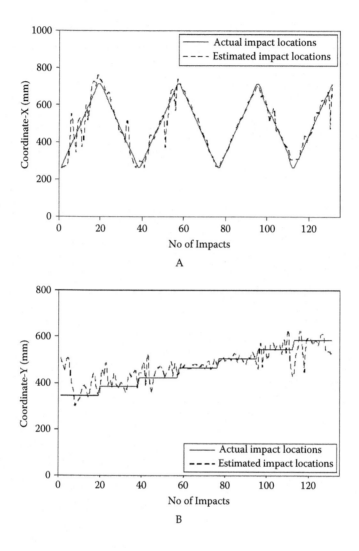

FIGURE 6.13
Comparison between the actual and estimated impact coordinates for zone 2 when all seven of the features are used: (A) *x* coordinate, (B) *y* coordinate.

TABLE 6.3

Local Impact Estimation Percentage Errors for the Composite Wing Box Structure Based on the ANN-Based Procedure

	Features	Zone 1	Zone 2	Zone 3	Zone 4
	Maximum peak	0.81	0.81	3.51	1.35
	Minimum peak	0.81	0.71	3.72	1.73
	Peak-to-peak	0.67	0.38	3.00	1.21
	Variance	1.87	1.57	4.30	2.04
	Arithmetic mean (absolute)	1.06	0.34	3.11	1.91
ANN-based	Arithmetic mean (real)	0.72	0.42	2.71	1.70
	Arithmetic mean (imaginary)	1.29	1.14	3.59	1.92
	Combined time domain	0.92	0.65	2.61	1.23
	Combined frequency domain	1.01	0.48	2.41	1.48
	All features	0.69	0.51	2.39	1.27

set (526 points) and a testing set (447 points). The most suitable number of hidden layers and hidden layer nodes was determined for an MLP with 15 hidden layer nodes as the optimum hidden node number for impact location (Figure 6.14). Finally, once the optimal network had been constructed and

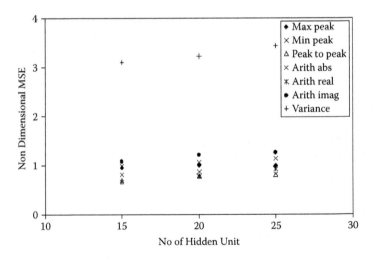

FIGURE 6.14
Hidden unit selections for the composite wing box structure (entire strain data signal features used for training).

validated, the testing data set was used to give the final estimated results, which were compared with the actual impact coordinates.

The global impact location results shown in Figures 6.15, 6.16, and 6.17 indicate the performances for time-domain features, frequency-domain features, and combinations of all signal features, respectively. The results show that the overall trend for the estimated coordinates is maintained if compared with the relevant actual coordinates for all parameters investigated.

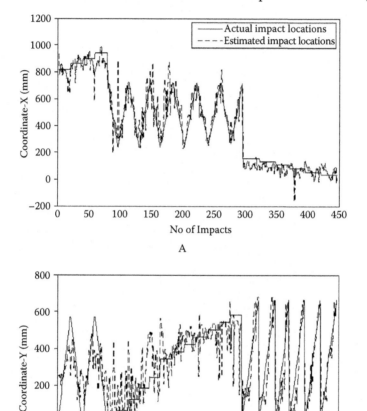

FIGURE 6.15
Comparison between the actual and estimated impact coordinates when the peak-to-peak signal feature is used: (A) x coordinate, (B) y coordinate. The analysis is based on data from the entire structure.

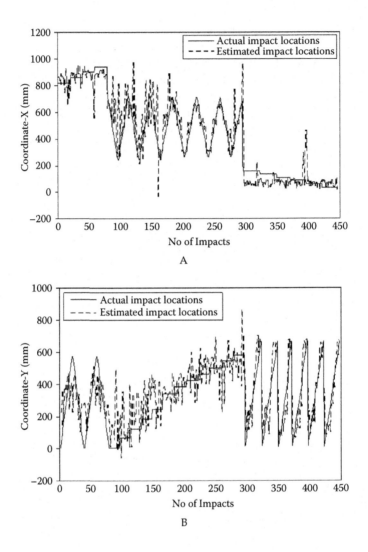

FIGURE 6.16
Comparison between the actual and estimated impact coordinates when the real spectrum arithmetic mean feature is used: (A) x coordinate, (B) y coordinate. The analysis is based on data from the entire structure.

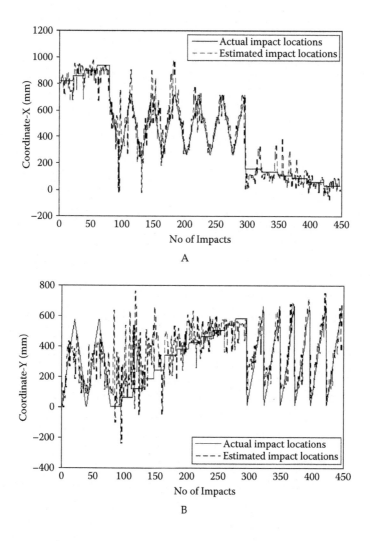

FIGURE 6.17
Comparison between the actual and estimated impact coordinates when all seven of the signal features are used: (A) x coordinate, (B) y coordinate. The analysis is based on data from the entire structure.

TABLE 6.4

Summary for the MLP Regression Results for the
Composite Wing Box Structure

Features	ANN-based (regression) [%]
Maximum peak	0.95
Minimum peak	1.03
Peak-to-peak	0.71
Variance	3.24
Arithmetic mean (absolute spectra)	0.86
Arithmetic mean (real spectra)	0.80
Arithmetic mean (imaginary spectra)	1.11
Combined time domain	0.87
Combined frequency domain	0.91
All features	0.83

The analysis is based on the data from the entire structure.

This shows that, in general, the ANNs were capable of estimating the global impact location. The impact estimation errors for all signal features used for a wing box structure are summarized in Table 6.4. The best feature in estimating the global impact location was the arithmetic mean (real spectra), whereas the worst feature was the variance. It also shows that the combination of all features together lead to good estimates of impact location.

6.10   Conclusions

This chapter presents the application of ANN with an MLP procedure to estimate impact location in a complex composite structure using impact strain data from a network of passive transducers. ANN with an MLP procedure can be applied to transform these data into meaningful location information, that is, impact coordinates and/or the structural areas where impacts have occurred. The findings show that ANNs are capable of estimating both local and global impact detection. The overall percentage error, in terms of the total structural area monitored, was less than 5%. The selection of features from impact strain data is important for impact location in the ANN-based procedure. Altogether, the peak-to-peak amplitude value from the time domain and the arithmetic mean from the real part of the frequency spectrum produced the best impact location results. Combining various features as network inputs does not necessarily lead to improved results, as this approach led to better impact location results for the smart composite plate but a worse network performance for the wing box structure. The geometry and material properties of monitored structures can affect impact location results. Investigations on the complex wing box structure illustrated that the best results were obtained for the central part of the structure, where less complexity was involved (i.e., a flat composite skin and no structural inserts). The leading and trailing edges of the wing box structure produced worse results due to curvatures of the composite skin and substantial wave attenuation in the honeycomb inserts. In summary, the impact location methods presented in this chapter have provided good results and offer great promise for application in real engineering structures.

References

Diamanti, K., Hodgkinson, J. M., and Soutis, C. 2004. Detection of low-velocity impact damage in composite plates using Lamb waves. *Structural Health Monitoring* 3:33–41.

Dupont, M., Osmont, D., Gouyon, R., and Balageas, D. L. 1999. Permanent monitoring of damaging impacts by a piezoelectric sensor based integrated system, in *Proceedings of the 2nd International Workshop on Structural Health Monitoring*. Stanford University, California. Technomic Publishing Co. Inc., 561–570.

Grondel, S., Assaad, J., Delebarre, C., and Moulin, E. 2004. Health monitoring of a composite wingbox structure. *Ultrasonics* 42:819–824.

Haywood, J., Coverley, P. T., Staszewski, W. J., and Worden, K. 2005. An automatic impact monitor for a composite panel employing smart sensor technology. *Smart Materials and Structures* 14:1–7.

Haywood, J., Staszewski, W. J., and Worden, K. 2001. Impact location in composite structures using smart sensor technology, and neural networks, in *Proceedings of the 3rd International Workshop on Structural Health Monitoring: The demands and challenges*. Stanford University: CRC Press, 1466–1475.

Kessler, S. S., Spearing, S. M., and Soutis, C. 2002. Damage detection in composite materials using Lamb wave methods. *Smart Materials and Structures* 11:269–278.

Kirikera, G. R., Shinde, V., Schulz, M. J., Ghosal, A., Sundaresan, M., and Allemang, R. 2007. Damage localisation in composite and metallic structures using a structural neural system and simulated acoustic emissions. *Mechanical Systems and Signal Processing* 21:280–297.

Le Clerc, J. R., Worden, K., Staszewski, W. J., and Haywood, J. 2004. Impact detection in an aircraft composite panel—a neural network approach, in *Proceeding of the 2nd European Workshop on Structural Health Monitoring*, Munich, Germany, 407–414.

Le Clerc, J. R., Worden, K., Staszewski, W. J., and Haywood, J. 2007. Impact detection in an aircraft composite panel—a neural-network approach. *Journal of Sound and Vibration* 299:672–682.

Lee, J. W., Kirikera, G. R., Kang, I., Schulz, M. J., and Shanov, V. N. 2006. Structural health monitoring using continuous sensors and neural network analysis. *Smart Materials and Structures* 15:1266–1274.

Mal, A., Ricci, F., Banerjee, S., and Shih, F. 2005. A conceptual structural health monitoring system based on vibration and wave propagation, *Structural Health Monitoring* 4:283–293.

Prasad, S. M., Balasubramaniam, K., and Krishnamurthy, C. V. 2004. Structural health monitoring of composite structures using Lamb wave tomography. *Smart Materials and Structures* 13:N73–N79.

Sekhar, B. V. S., Balasubramaniam, K., and Krishnamurthy, C. V. 2006. Structural health monitoring of fiber-reinforced composite plates for low-velocity impact damage using ultrasonic Lamb wave tomography. *Structural Health Monitoring* 5:243–253.

Staszewski, W. J., Biemans, C., Boller, C., and Tomlinson, G. R. 2002. Impact damage detection in composite structures using passive acousto-ultrasonic sensors. *Key Engineering Materials* 221–222:389–400.

Staszewski, W. J., Worden, K., Wardle, R., and Tomlinson, G. R. 2000. Fail-safe sensor distributions for impact detection in composite materials. *Smart Materials and Structures* 9:298–303.

Sung, D., Hoon, J., Kim, C. G., and Hong, C. S. 2000. Impact monitoring of smart composite laminates using neural network and wavelet analysis. *Journal of Intelligent Material Systems and Structures* 11:180–190.

Tarassenko, L. 1998. *A guide to neural computing applications*. London: Arnold.

Toyama, N., and Takatsubo, J. 2004. Lamb wave method for quick inspection of impact-induced delamination in composite laminates. *Composites Science and Technology* 64:1293–1300.

Walz, M. 2006. The dream of composites, in R&D, features—Advantage Business Media.

Wang, C. S., and Chang, F. K. 1999. Built-in diagnostic for impact damage identification of composite structures, in *Proceedings of the 2nd International Workshop on Structural Health Monitoring*, Stanford University, California. Technomic Publishing Co. Inc., 612–621.

Wolfinger, C., Arendts, F. J., Friedrich, K., and Drechsler, K. 1998. Health-monitoring-system based on piezoelectric transducers. *Aerospace Science and Technology* 2(6):391–400.

Worden, K. 1996. *MLP-Multi-layer perceptron, Version 3.4—user manual*, ed. U.o.S. Dynamic Research Group.

Worden, K., and Staszewski, W. J. 2000. Impact location and quantification on a composite panel using neural networks and a genetic algorithm. *Strain* 36:61–70.

Yang, Y. C., and Han, K. S. 2002. Damage monitoring and impact detection using optical fiber vibration sensors. *Smart Materials and Structures* 11:337–345.

7

Artificial Neural Networks for Predicting the Mechanical Behavior of Cement-Based Composites after 100 Cycles of Aging

E. M. Bezerra, C. A. R. Brito Jr., A. C. Ancelotti Jr., and L. C. Pardini

CONTENTS

ABSTRACT   In this chapter, an artificial neural network model was created to predict the bending stress-strain behavior from cement-based composites after accelerated aging test (100 cycles), considering the following effects: the volumetric concentration of fibers (V_f), fiber specimen (F_s), and age of testing (A_t). To reach this goal, a multilayered neural network perceptron architecture was used, and results show that using this architecture plus the conjugate gradient (CG) algorithm leads to a high predictive quality to cement composites, that is, nearly 70% of standard error of prediction was found to be ≥ 0.9. The initial tests considered a simple architecture, 3-[10-10]$_2$-2, resulting in low predictive quality. However, increasing the number of neurons in the hidden layers and the number of training instances resulted in an enhancement of the neural network predictive quality. It was verified that each one of the parameters quoted was responsible by a certain level of impact.

7.1 Introduction

In the early 1970s, a global effort was initiated to legislate for the removal of asbestos reinforcement from a wide range of products (Coutts, 2005). Australia was the first country to be totally free of asbestos in fiber cement production. James Hardie Industries have been manufacturing asbestos-free cement sheeting since 1981. In short, the technology developed for James Hardie Industries consists of autoclaving its sheets using cellulose fibers as reinforcement. Nevertheless, these fibers show ineffectiveness, compared to asbestos, in trapping the cement particles during formation of the sheet on the conventional Hatschek machine. In addition, when cellulose fiber-reinforced cement composites were exposed outdoors, composite performance and mechanical properties changed, implying poor durability (Kim et al., 1999). This process is related to a degradation of the fiber by the alkaline cement paste environment and the increase in fiber dimensions related to variations in humidity (Juárez, 2007). Agopyan et al. (2005) reported that the main disadvantage of the use of vegetable fiber is the durability of these fibers in a cementitious matrix and also the compatibility between both phases. The alkaline media weakens most natural fibers due to several chemical mechanisms. This can be partly attributed to the embrittlement that the cellulose fiber suffers due to its mineralization inside the cement matrix (Akers and Studinka, 1989). The problem associated with the loss of mechanical properties after accelerated aging test can be solved by the use of synthetic fibers, which are less sensitive to alkali attack and possess good mechanical properties, resulting in composites with better performance in the long term (Bezerra, 2005). Polymeric fibers are increasingly being used for the reinforcement of cementitious matrix to improve the toughness and energy absorption capacity, and to reduce the cracking sensitivity of the matrix (Singh et al., 2004). The cement-based composites exhibit the general characteristics of brittle matrix composites, that is, the failure of the matrix precedes the fiber failure. The toughening of the matrix can be related to several mechanisms: fiber bridging, fiber debonding, fiber pullout (sliding), and fiber rupture as a crack propagates across a fiber through the matrix (Zollo, 1997). On the other hand, these fibers are effectively an order-of-magnitude greater in diameter than wood fibers and thus cannot by themselves form films and trap the particles. These fibers are more likely to settle out in the Hatschek machine, making the formation of the sheets difficult (Coutts, 2005). Thus, the no-asbestos cement-based composites should be reinforced with two types of fibers: cellulose fibers (used as process fiber) and synthetic fibers (used as the main reinforcement after aging).

Therefore, evaluating the mechanical properties of these composites is a complex task. Bezerra and Savastano (2004) used regression analysis to evaluate six different variables related to formulation and processing and their effect on the mechanical and physical behavior of nonconventional fiber cements. The conclusions obtained using this technique only contributed to indicate which are the main parameters (Bezerra and Savastano, 2004). Thus, there arises a need for using more elaborate nonlinear algorithms such as artificial neural networks (ANNs) to refine the prediction of the mechanical properties of cement-based composites.

Evaluating the mechanical behavior of the specimens after cycles of accelerated aging test is of importance in the design and inspection of cement-based composites such as water tanks and roofing (Bezerra et al., 2006). In the absence of a well-defined failure criterion that can be used to predict stress bending, extensive tests must be performed for different specimens of fibers, volumetric concentrations of reinforcement, and cycles of aging (Bezerra et al., 2004). The composite material structure is subdivided into three areas: fiber, matrix, and fiber-matrix interfacial transition zone. The micromechanical parameters that should be used to determine the material structure are fiber parameters for strength, stiffness, length, and diameter; matrix parameters for stiffness, fracture toughness, and initial flaw size; and fiber-matrix interaction parameters for interfacial frictional and chemical bond and snubbing coefficient. These input parameters can be used in a flexural micromechanics-based model to predict the composite mechanical properties (Maggio et al., 1997). The durability of cellulose composites is theoretically assessed by recognizing the influence of a changing environment (dry, wet, or aged) on fiber, matrix, and interfacial properties, or as a whole, the stress-strain relation in flexure. These changes influence the composite's flexural mechanical properties and their load deflection behavior. Kim et al. (1999) applied the micromechanics-based durability model to predict the flexural mechanical properties of thin sheet cementitious composites reinforced with refined and unrefined cellulose fibers when the composites were tested at dry and wet conditions to verify the mechanical behavior after accelerated aging tests. This model correctly predicted the trends in composite properties [limit of proportionality, modulus of rupture (MOR)] with changing environment, yet there is a discrepancy when predicting composite toughness for dry and wet conditions. It has been recognized that because of the heterogeneity inherent in the microstructure of concrete, strain softening, microcracking, and larger-scale process zone, in the order of meter, fracture parameters such as fracture toughness (K_{IC}) and fracture energy (G_f) determined in accordance with linear elastic fracture mechanics are size dependent (Zhang and Li, 2004).

In this manner, the micromechanical models present severe limitations in the prediction of cement-based composites, which makes its application very difficult. Bearing in mind the limitations of the phenomenological model, an alternative approach is to use different computational and knowledge representation paradigm, that is, neural networks. A potential solution of this requirement is offered by ANNs. ANNs are an alternative to conventional programmed computing and are based on the operation of the brain. An ANN mimics the structure and functionality of a biological nervous system, and it runs a parallel distributed processing mode of computation. It is capable of making decisions based on incomplete, noisy, and disordered information, and can generalize rules from those cases in which it was trained, applying these rules to new stimuli. A neural network architecture is a promising implicit modeling scheme based on learning a set of parameters (weights), aimed at replacing the traditional explicit constitutive equations used to describe material behavior (Zhang et al., 2002, 2003; Huber and Tsakmakis, 2001; Al-Haik et al., 2004; Al-Assaf and El Kadi, 2001). In a previous work, Bezerra et al. (2007) concluded that ANNs can lead to high predictive quality for the prediction of the shear mechanical behavior of epoxy matrix composites reinforced with carbon and glass fibers for two different angle of orientations layer: $0°/90°$ and $\pm 45°$.

This chapter introduces an ANN model to predict the bending mechanical performance of cementitious composites considering the following aspects: (1) effect of the incorporation of different types of synthetic fibers in cementitious matrix; (2) the fiber specimen; and (3) aging at different curing dates. The stress-strain curves are typically nonlinear because they are a matrix-dominated property in which interfacial properties play a key role. The design and training of the neural networks were performed considering a multilayer feed-forward architecture with supervised learning through a CG algorithm (see Section 7.2), implemented using the Neural Network Toolbox from MATLAB (Demuth and Beale, 2000) and using as training data the results obtained after 28 days of cure and 50 cycles of aging. The results obtained after 100 cycles were used only to test and validate the neural network.

7.2   Conjugate Gradient Algorithm

The CG algorithm (Abraham, 2004; Kalman and Kwasny, 1997) was recently introduced as a learning tool in neural networks because of the following reasons: it is quadratically convergent, it behaves as well as backpropagation in tough regions, and it is possible to use a less expensive derivate-free line search for most of the training. This method requires linear storage (in weights), whereas Newton-style quadratically convergent methods require quadratic storage. Furthermore, the CG algorithm uses fewer epochs compared to backpropagation algorithm, but maintains generalization properties almost similar to those of backpropagation techniques. This algorithm provides a tool to update the learning rate to minimize the square error mean square error (MSE); hence, the learning rate is updated at each epoch. The scale conjugate gradient (SCG) version of the CG algorithm was used in this work. This variation of the CG algorithm was developed by Moller (1993) to avoid the time-consuming line search.

Within a neural network context, CG methods can be thought of as extensions of the steepest descent to include variable learning rate. The estimation procedure begins with the calculation of the first derivative of the error function (E) in relation to the initial weight (w_0):

$$s_0 = -\frac{\partial E}{\partial w_0} = -g_0 \tag{7.1}$$

The minimization procedure of the error function E led to the generation of a series of approximators (w_x). For $x = 0, 1$, start at point w_x, then perform a line minimization of E in the direction s_x. This method is equivalent to determining a scalar λ to minimize the function expressed by:

$$J(\alpha) = E(w_x + \lambda s_x) \tag{7.2}$$

J is expressed as a function of the learning rate λ for fixed values of w and s. The next estimation of w is given by:

$$w_{x+1} = w_x + \lambda s_x \tag{7.3}$$

A version of the CG algorithm that does not require line searches was developed by Moller (1993) and uses the finite difference method for estimating Hd_k. The parameter δ is defined to monitor the sign of the product $d_k{}^T Hd_k$ according to Equation (7.4).

$$\delta = d_k^T Hd_k \tag{7.4}$$

Furthermore, Moller introduced two new parameters, λ and $\bar{\lambda}$, to define an altered value of δ, $\bar{\delta}$. These parameters are altered, ensuring that $\bar{\delta} > 0$. This algorithm produces a step size that shows good results in practice. $\bar{\delta} > 0$ is defined as:

$$\bar{\delta} = \delta + (\bar{\lambda} - \lambda)d_k^T d_k \tag{7.5}$$

The requirement for $\bar{\delta} > 0$ gives a condition for $\bar{\lambda}$:

$$\bar{\lambda} > \lambda - \frac{\delta}{d_K^T d_K} \tag{7.6}$$

In this method, Moller defines $\bar{\lambda} = 2(\lambda - (\delta / d_K^T d_K))$ to satisfy Equation (7.6) and so ensures that $\bar{\delta} > 0$. Equation (7.7) shows the results of substituting this expression in Equation (7.5):

$$\bar{\delta} = -\delta + \lambda d_K^T d_K \qquad (7.7)$$

This procedure is detailed step by step by Moller (1993).

7.3 Experimental Work

7.3.1 Composite Production

Hardened specimens of cement-based composites were produced with dimensions of 40 × 160 mm, ~5 mm thickness, and reinforced with polyvinyl alcohol (PVA), polyacrylonitrile (PAN), and polypropylene (PP) fibers, which are described in Table 7.1. The matrix was composed of ordinary Portland cement CPII F (specified by Brazilian Standards NBR 11578) (Associação Brasileira de Normas Técnicas, 1991). The chemical composition of this cement is given in Table 7.2. The specific surface area (BET technique) of the CPII F cement is 598 m²/kg according to the information provided by the manufacturers. Carbonate filler with specific area of 451 m²/kg was used as an aggregate. The characteristics of the silica fume Elkem-920D type are described in Table 7.3 and the specific surface area is 22.5 m²/g. One type of cellulose pulp was used to assist with filtering in the fiber cement production and reinforcement in the hardened composite: Brazilian *Pinus taeda* unbleached kraft pulp with Kappa number of ~45, and °SR = 65. This capability is of particular interest for the retention of matrix constituents during the production of composite slates in industrial scale by the Hatschek (or wet) process (Coutts, 1988). Table 7.4 shows all formulations used and their respective percentage by volume of solid raw materials. The specimens were produced in the laboratory by slurrying the raw material in water solution containing 20% of solids,

TABLE 7.1

Geometric and Mechanical Properties of Synthetic Fibers

Sample	Length (µm)	Diameter (µm)	Aspect Ratio	Stiffness (GPa)	Density (g cm⁻³)
PVA	6000	14	429	19.8	1.30
PP	5600	26	215	1.3	0.92
PAN	6000	12	500	17.7	1.19

TABLE 7.2

Oxide Composition of the Ordinary Portland CPIIF Cement

	Content (% by mass)
Compound	CPIIF*
Ignition loss	3.69
SiO_2	19.53
CaO	60.49
Al_2O_3	4.71
Fe_2O_3	3.07
MgO	4.68
Na_2O	0.10
K_2O	0.96
SO_3	2.60
CO_2	—
Insoluble residue	—

Source: Laboratory of Microstructure/PCC, Escola Politécnica, USP, Brazil.

TABLE 7.3

Physical and Chemical Properties of the Pozzolanic Materials

Material	Pozzolanic Activity[a] (mg/g)	BET (m²/g)	Chemical Composition (%)			
			SiO_2	Al_2O_3	CaO	Ignition Loss
Silica fume	813.83	22.50	91.2	0.22	0.22	1.30

Source: Laboratory of Microstructure/PCC, Escola Politécnica, USP, Brazil.
[a] Based on the Chapelle method.

TABLE 7.4

Formulations of Composites (% by Volume of Solid Raw Materials)

Formulation	Synthetic Fiber	Brazilian Softwood Pulp 65°SR	Cement	Carbonate Filler	Silica Fume
PVA/PP/PAN 1.8	1.8	9.4	68.6	13.8	6.4
PVA/PP/PAN 2.8	2.8	9.4	68.6	13.8	5.4
PVA/PP/PAN 3.2	3.2	9.4	68.6	13.8	5.0
PVA/PP/PAN 3.7	3.7	9.4	68.6	13.8	4.5

followed by a vacuum drainage of excess water and pressing at 3.2 MPa for 300 s. This preparation procedure is described in detail elsewhere (Savastano et al., 2001). The final ratio between water and solids at the end of slurry dewatering and pressing processes can only be estimated, and it is expected to be about 0.5 for the amount of fibers used (Table 7.4) as discussed by Bezerra (2005) and Caldas e Silva (2002). Ten specimens of each formulation were subjected to wet curing for 7 days, followed by air curing until lasting 28 days when the mechanical properties were assessed. Twenty specimens for each formulation were subjected to the accelerated aging test (soak and dry cycles), of which half were subjected to 50 cycles and the rest to 100 cycles. This test consists of submerging the specimens in cold water for 18 h. Next, they were placed in an oven set at 60°C for 6 h, to complete the cycle of 24 h. The aging test is composed by 50 cycles and it is based on the methodology of the European Standards EN-494 section 7.3.5 (Caldas e Silva, 2002; European Committee for Standardization, 1994). The 100 cycles of aging is a special condition to verify the mechanical behavior of cementitious matrix composites with higher exposition to degradation.

7.3.2   Composite Characterization

The mechanical behavior was based on the Rilem recommendations 49 TFR (CEN, 1994). It was assessed by a four-point bending configuration. The loading apparatus consisted of two supports a distance $L = 135$ mm apart and two loading points spaced to provide third point loading ($L/3$). A deflection rate of 1.5 mm/min was used for all tests in an Emic DL30000 universal testing machine equipped with a load cell of 1 kN (Savastano et al., 2001). The mechanical properties evaluated at 28 days and after accelerated aging test were MOR and toughness.

7.3.3   Proposed Model

For bending prediction analysis of PVA, PP, and PAN fiber composites, it was assumed that stress bending is a function of three parameters: volumetric concentration of fibers (V_f), specimen of fiber (S_f), and age of testing (A_t). Furthermore, all samples have similar geometric features and precautions were taken to make sure that the manufacturing and testing conditions were similar in all cases. The ANN model proposed for predicting the bending stress-strain behavior of a fiber cementitious matrix composite is illustrated in Figure 7.1. It consists of a number of simple neuron-like processing elements, also called units or nodes, organized in layers that are classified as input layer, hidden layer(s), and output layer. These connections are not all equal; each connection has neurons with different weight and associated bias. These classifiers adjust internal parameters

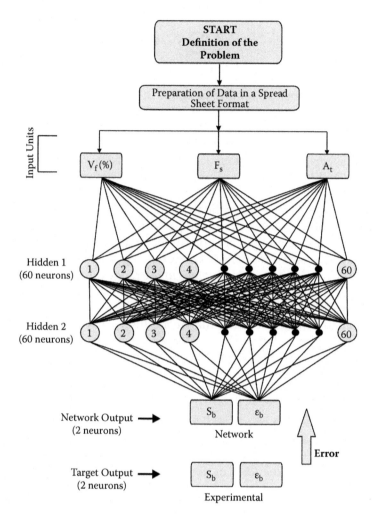

FIGURE 7.1
Architecture of a feedforward back propagation neural network.

W performing vector mappings from the input to the output space $Y^{(p)}$ $= A_W(X^{(p)})$. In this way, the data were processed at the input layer and followed by network structure constituted by two hidden layers until it arrives at the output layer.

Each neuron has an associated bias (a neuron threshold link with a unitary input) to adjust its output. The output layer contains two neurons, whose output is the estimated stress-strain curve. The objective of the training procedure is to find a set of possible weights, w_{ij}, that will enable the network to produce the prediction (Y_p) as similar as possible to the known experimental output (Y_e). This is achieved by minimizing the cost function E, which is the MSE, as shown below:

$$E = \frac{1}{2q} \sum_{i}^{q} (Y_e - Y_p)^2 \tag{7.8}$$

where q is the number of training patterns. In this work, we used the back-propagation algorithm associated with the CG algorithm as described in Section 7.2.

Because we have three inputs and two outputs, three neurons suffice for the input layer and two neurons for the output layer. Training data were normalized to be in binary form for speeding up network training. About 70% of the data has been used in the training, and simulation steps and the rest of the data were used for validation and testing of the network model. The structure of the network can be represented as [3, HL1, HL2, 2], where HL1 and HL2 are the number of nodes in the first and second hidden layers, respectively. We started with the network topology [3,10,10,2], initializing all the weights by assigning random values in the range (−1, +1), using the training data to learn the weights, and recording the value of MSE. We then grew the network up to [3,60,60,2] by adding five neurons at a time to the hidden layer and then similarly treating the second hidden layer. We repeated the training procedure each time the network topology was modified. This process is continued until a prescribed MSE error tolerance of 10^{-4} is satisfied.

7.3.4   Evaluation of the ANN Performance

A data set of measured results will usually be divided into three data sets: training, testing, and validation of the neural network. The training data set is used to adjust the weights of all the connecting nodes until the desired error level is reached. The ANN performance can be evaluated using the coefficient of determination B (also called R^2 coefficient), which is defined by:

$$B = 1 - \frac{\sum_{i=1}^{M} (O(p^{(i)}) - O^{(i)})^2}{\sum_{i=1}^{M} (O^{(i)} - O)^2} \tag{7.9}$$

where $O(p^{(i)})$ is the ith predicted property characteristic, $O^{(i)}$ is the ith measured value, O is the mean value of $O^{(i)}$, and M is the number of test data. The coefficient B describes the fitness of the ANN output variable approximation curve to the actual test data output variable curve. Higher B coefficients indicate an ANN with better output approximation capabilities (Rillem, 1984).

To avoid any influence in selecting the test data, a random technique was applied in the selection, and the entire process is repeated independently many times (e.g., 50 times). Afterward, the distribution of B values is recorded and the percentage of $B \geq 0.9$ is calculated, since this value is identified as corresponding to a high predictive quality, that is, less than 15% of the root mean square error is between the predicted values and the measured ones. It is clear that the higher the percentage of B ($B \geq 0.9$), the better the quality (Bezerra et al., 2007). Another aspect that should be observed is the increase of the percentage of test data with a B value of ≥ 0.9 as a function of the number of neurons in the hidden layers.

7.4 Results and Discussions

The number of hidden nodes in the second and third layers was varied between 10 and 60 hidden units. Each individual network structure was trained with 500 independent training sessions, starting at differently randomly chosen points on the error surface so that a total 11 independent networks were evaluated. For each of the 10 to 60 hidden node network structures, the error was calculated for the training, validation, and test data sets. These errors are tabulated in Table 7.5. According to the results, the topology $3\text{-}[60\text{-}60]_2\text{-}2$ displayed the best performance.

In this section, results are discussed using the network (Figure 7.1) to predict the MOR and toughness of cementitious composites reinforced with synthetic fibers after 100 cycles of aging. The maximum performance was reached using a multilayer perceptron composed by 60 neurons in two hidden layers.

This high number of neurons is understandable once the curves present highly nonlinear behavior, as shown in Figures 7.2, 7.3, and 7.4.

TABLE 7.5

Performance on the Entire Data Set using Different Topologies

Topologies	$3\text{-}[10]_2\text{-}2$	$3\text{-}[15]_2\text{-}2$	$3\text{-}[20]_2\text{-}2$	$3\text{-}[25]_2\text{-}2$	$3\text{-}[30]_2\text{-}2$
<0.6	33.76	26.69	23.35	22.79	21.81
[0.6–0.7]	2.88	15.12	10.92	9.84	8.86
[0.7–0.8]	1.79	12.17	7.78	7.41	7.06
[0.8–0.9]	8.45	9.45	8.20	7.96	7.73
[0.9–1.0]	53.12	36.57	49.74	52.00	54.53
Topologies	$3\text{-}[35]_2\text{-}2$	$3\text{-}[40]_2\text{-}2$	$3\text{-}[45]_2\text{-}2$	$3\text{-}[55]_2\text{-}2$	$3\text{-}[60]_2\text{-}2$
<0.6	11.18	7.18	6.92	4.38	0.24
[0.6–0.7]	7.99	8.19	6.48	5.84	4.78
[0.7–0.8]	6.72	7.40	6.10	5.81	5.98
[0.8–0.9]	7.51	7.29	7.08	6.87	10.36
[0.9–1.0]	66.60	69.94	73.43	77.10	78.64

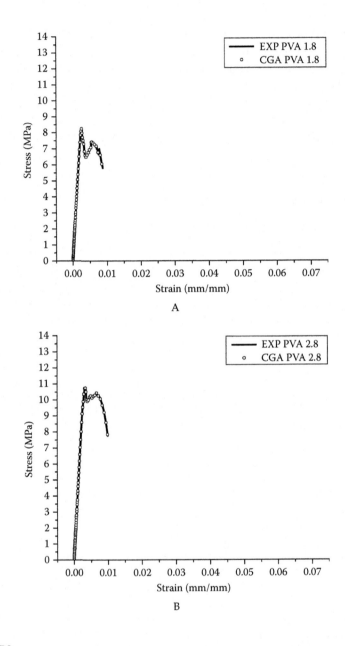

FIGURE 7.2
Stress-strain curves of the composite reinforced with different concentrations of PVA fibers after 100 cycles of accelerated aging test (in % by volume of raw materials).

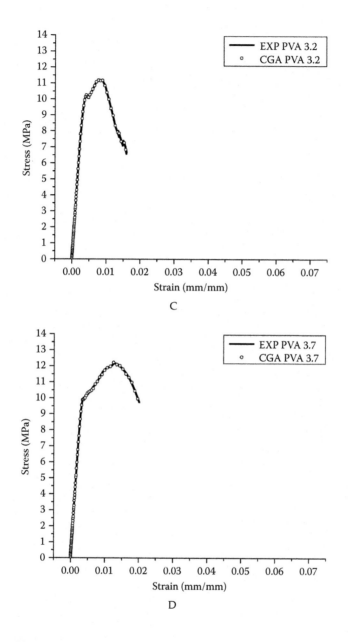

FIGURE 7.2 (CONTINUED)

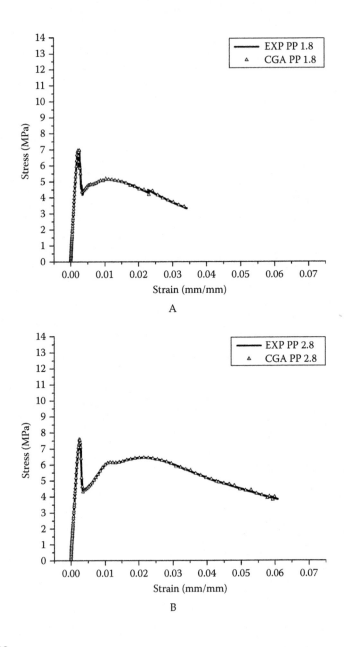

FIGURE 7.3
Stress-strain curves of the composite reinforced with different concentrations of PP fibers after 100 cycles of accelerated aging test (in % by volume of raw materials).

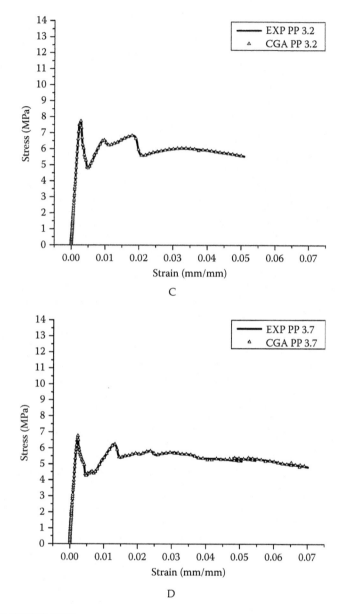

FIGURE 7.3 (CONTINUED)

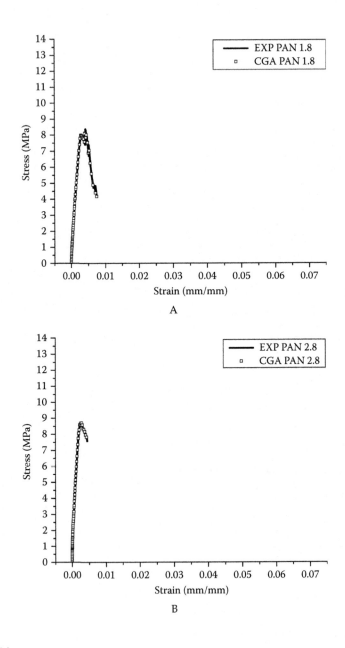

FIGURE 7.4
Stress-strain curves of the composite reinforced with different concentrations of PAN fibers after 100 cycles of accelerated aging test (in % by volume of raw materials).

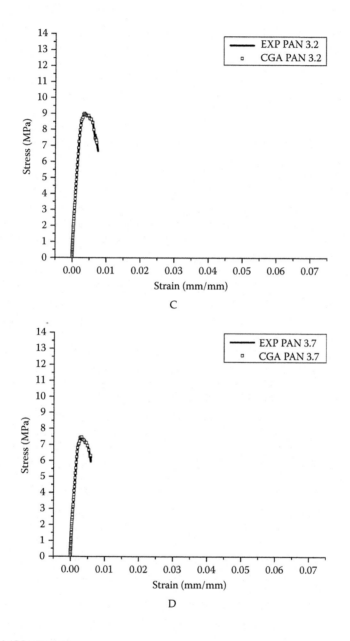

FIGURE 7.4 (CONTINUED)

Because of the high speed of convergence and low storage, and considering that this network and the input set are large, we chose CG to train the neural network, although it was necessary to have more than 2500 epochs to reach an MSE error of less than 10^{-4} and more than 3000 epochs to reach the equilibrium position (in other words, a point where the error in the validation test reaches a minimum). The average sum squared error plotted as a function of the training, validation, and testing cycles is shown in Figure 7.5 for the neural network with 60 hidden neurons. Experiments indicated that there was significant improvement in convergence as the number of hidden neurons increased beyond 10. The ANN predictions for the training and testing sets are shown in Figure 7.6a and b. The scattered nature of the predicted values vs. the measured values was assessed using regression analysis. Results show that the ANN was successful in modeling the nonlinear relationship between the experimental conditions and the parameters used. This is also evident in Figures 7.2, 7.3, and 7.4.

Moreover, this algorithm was selected for several reasons: it is quadratically convergent, it behaves as well as backpropagation, and it is possible to use less expensive derivative free line search for most of the training (Kalman and Kwasny, 1997). Furthermore, the application of the CG algorithm makes the predicted curves similar to those experimentally determined after 100 cycles of the accelerated aging test. This progress is important, because it will enable the handlers to predict mechanical behavior for other curing dates without having to perform new experiments, translating to cost and time savings when generating new results. The optimum micromechanical models used to predict the mechanical properties of composites consider that the fibers are homogeneously dispersed in the matrix (Haykin, 1998). In this fashion, in most cases the calculated results applying to these models are not comparable to those obtained by experimentation. The major contribution of neural networks is in predicting the behavior of cementitious matrix after 100 cycles of aging.

In spite of the large number of epochs required to reach the goal, this network has proven to be a good model for adjusting the stress-strain curves of cement-based composites, once it achieves an excellent quality of prediction as shown in the graphics above. In all curves, the output of network almost exactly follows the experimental curves. As shown in Figure 7.7, the neural network performance is associated with the robustness of its performance. Note that ANN 3-[60-60]$_2$-1 presented superior performance to ANN 4-[10-10]$_2$-1. The results showed that application of the CG algorithm leads to a high predictive quality, that is, nearly 95% of B was found to be ≥0.90 for the best ANN configuration.

The first parameter (volumetric concentration of fibers) significantly affected the toughness of composites reinforced with a smaller fraction of reinforcement. The formulations containing a smaller concentration of synthetic fibers suffered higher degradation after 100 cycles. This behavior can be attributed to the degradation of the composites, since the synthetic

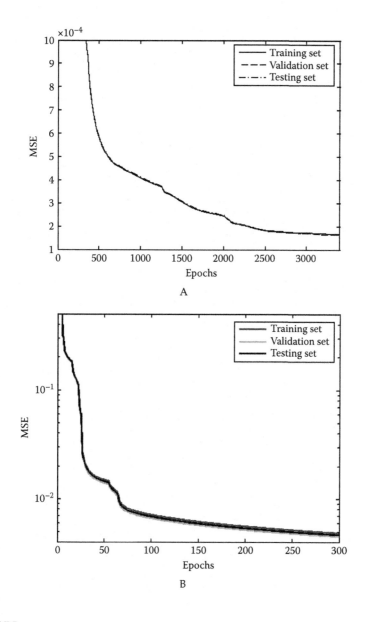

FIGURE 7.5
MSE error for training, validation, and testing sets for the 4-[60-60]2-1 ANN with backpropagation training algorithm that uses conjugate gradient algorithm.

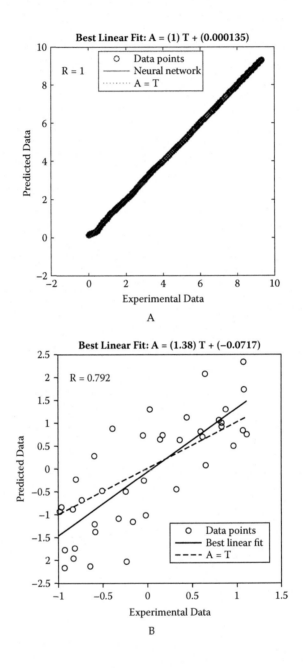

FIGURE 7.6
Dependence of the percentage of test data on *B* value as a function of the network architecture for cement matrix composites reinforced with synthetic fibers.

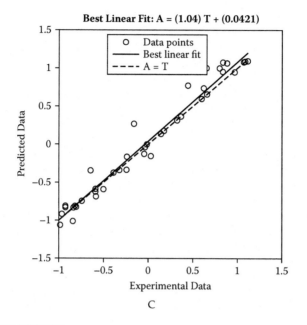

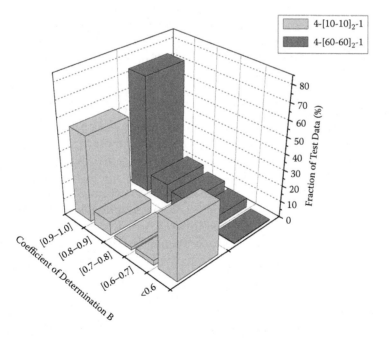

FIGURE 7.7
Dependence of the percentage of test data on *B* value as a function of the network architecture for cementitious matrix composites.

fibers are present in proportionally smaller amounts in these formulations, and the cellulose fibers are sensitive to alkali attack (Coutts, 2005). The synthetic fibers are the major factor responsible for the mechanical resistance of composites, whereas the secondary role is ascribed to cellulose fibers. Moreover, cellulose fibers are important to retain cement particles in industrial processes and some reinforcement effects in the early stages (Bezerra et al., 2007; Guzmán de Villoria and Miravete, 2007). The peeling and the hydrolytic reactions (Bezerra et al., 2006) are responsible for the loss of polysaccharides and reduction of the cellulose chain length in alkaline environment (pH ~12–13). As shown in Figures 7.2, 7.3, and 7.4a, formulations reinforced with 1.8% (by volume) of synthetic fibers were more affected after aging than formulations reinforced with higher volumetric concentrations.

In relation to the second parameter (specimen of fiber), it was verified that the composites reinforced with PVA presented a higher MOR than formulations reinforced with PAN or PP after 100 cycles. This can be attributed to the hydroxyl groups on the PVA fiber surface, which cause an increase in the wettability of the fiber in the polar matrix, enhancing the dispersion of fibers when coupled with mechanical agitation on mixing (Bezerra, 2005). In this manner, the adhesion between the PVA fibers and cementitious matrix improves with the aging, being responsible for efficiency of load transfer. Otherwise, the cement-based composites reinforced with PP fibers presented higher toughness. These fibers present smoother pullout surface than the composite reinforced with PVA fibers, resulting in poorer anchorage in the cementitious matrix. PP fibers are easily pulled out during the load application. As a consequence, an increase in toughness and elongation at break has occurred with the reduction of the MOR of the PP composites when compared to those with PVA or PAN. This behavior can also be attributed to the following factors: geometric characteristics of these fibers, lesser surface rugosity, and aspect ratio in comparison with the other fibers used (Bezerra et al., 2004).

The PAN fibers anchored in the cementitious matrix after accelerated aging test (Beaudoin, 1990). This fact supports the assumption of fibers' brittle failure after aging due to an increase in bond strength of these fibers to the matrix after additional curing. The reduction in toughness of PAN cementitious composites should be related to the bonds formed with the cement matrix and the highest aspect ratio (~500) of this fiber. The highest aspect ratio of these fibers can make dispersion difficult, causing the inclusion of nonpermeable pore spaces in the composite in a variable manner, which can interfere with the test results. The adsorption of hydrophilic groups on PAN surfaces improves the adhesion between the cement matrix and the PAN fibers, which makes the pullout of these fibers difficult (Guzmán de Villoria and Miravete, 2007).

7.5 Conclusions

Formulations containing PVA fibers presented superior MOR than formulations with the same volumetric concentration of other synthetic fibers after 100 cycles of aging. This result can be associated to several factors: dispersion, anchoring, aspect ratio, fiber/matrix interface, and geometric characteristics.

Formulations reinforced with PP fibers presented superior toughness compared to other formulations because the factors cited above had favored the pullout of these fibers during the application of load. Indeed, PVA and PAN presented better interaction with the cementitious matrix, which made pullout difficult.

The use of ANNs led to high predictive quality for the prediction of the stress-strain curves of cementitious matrix reinforced with synthetic fibers after 100 cycles of accelerated aging test. This conclusion is supported by the results illustrated in Figures 7.2, 7.3, and 7.4. It was observed that for one data set exactly the size of the training data, the ANN with a superior number of neurons in the hidden layers showed a performance coefficient (B) superior to the ANN of lesser robustness. Therefore, it was observed that the architecture of ANN had an influence in the prediction quality of the bending profile of the cement matrix composites.

For future work, we are considering working out the mechanisms for extracting explanatory rules in ANN models, in such a way that possible hidden relationships among variables can be obtained. We are also considering the so-called neurofuzzy architectures that naturally allow a linguistic interpretation of the model, which can help both as an interpretatory mechanism for the complex nonlinearities expressed by the network and as a prior knowledge for building up more consistent phenomenological models in the prediction of cement-based composites.

References

Abraham, A. 2004. Meta learning evolutionary artificial neural networks. *Neurocomputing* 56:1–38.

Accion, F., Gobantes, J., and Blanco, M. T. 1990. Cements reinforced by acrylic fibers. Infrared studies: I. Hydration and hydrolysis processes in the fibers. *Cement Concrete Research* 20:702–710.

Agopyan, V., Savastano, H., Jr., John, V. M., and Cincotto, M. A. 2005. Developments on vegetable fiber-cement based materials in São Paulo, Brazil: an overview. *Cement Concrete Composites* 27:527–536.

Akers, S. A. S., and Studinka, J. B. 1989. Ageing behavior of cellulose fiber cement composites in natural weathering and accelerated tests. *Cement Concrete Composites* 11:93–97.

Al-Assaf, Y., and El Kadi, H. 2001. Fatigue life prediction of unidirectional glass fiber/epoxy composite laminate using neural networks. *Composite Structures* 53:65–71.

Al-Haik, M. S., Garmestani, H., and Savran, A. 2004. Explicit and implicit viscoplastic models for polymeric composite. *International Journal of Plasticity* 20:1875–1907.

Associação Brasileira de Normas Técnicas (ABNT). 1991. NBR 11578—Cimento Portland Composto. Rio de Janeiro [in Portuguese].

Beaudoin, J. J. 1990. *Handbook of fiber-reinforced concrete: principles, properties, developments and applications.* Park Ridge, NJ: Noyes.

Bezerra, E. M. 2005. Compósitos de matriz cimento Portland reforçados com fibras sintéticas de PVA/PP/PAN. PhD thesis, Aeronautic Institute of Technology, São José dos Campos, 123 pp. [in Portuguese].

Bezerra, E. M., Ancelotti, A. C., Pardini, L. C., Rocco, J. A. F. F., and Ribeiro, C. H. C. 2007. Artificial neural networks applied to epoxy composites reinforced with carbon and E-glass fibers: analysis of the shear mechanical properties. *Materials Science and Engineering A* 464:177–185.

Bezerra, E. M., Joaquim, A. P., and Savastano, H., Jr. 2004. Some properties of fiber-cement composites with selected fibers. In *Anais/NOCMAT*, ed. Ghavamy, K., Savastano, H., Jr., Joaquim, A. P., Anais da Conferência Brasileira de Materiais e Tecnologias Não-Convencionais: Habitações e Infra-Estrutura de Interesse Social Brasil-NOCMAT. Pirassununga(SP): Universidade de São Paulo; p. 12.

Bezerra, E. M., Joaquim, A. P., Savastano, H., Jr., John, V. M., and Agopyan, V. 2006. The effect of different mineral additions and synthetic fiber contents on properties of cement based composites. *Cement Concrete Composites* 28:555–563.

Bezerra, E. M., and Savastano, H., Jr. 2004. The influence of type and refinement of the cellulose pulp in the behavior of fiber cement with hybrid reinforcement—a regression analysis application. In *1st Symposium on ecologically friendly materials*, ed. Kaliakin, V. N., Kirby, J. T., Yamamuro, J., Bhattacharya, B. B., and Shenton, H. W. III, Proceedings 17th ASCE Engineering Mechanics Conference. Newark (DE): University of Delaware, pp. 1–8.

Caldas e Silva, A. 2002. Estudo da durabilidade de compósitos reforçados com fibras de celulose. Master dissertation, Universidade de São Paulo, São Paulo, 128 pp. [in Portuguese].

Coutts, R. S. P. 1988. Wood fiber reinforced cement composites. In *Natural fiber reinforced cement and concrete*, ed. Coutts, R. S. P., and Swamy, R. N., pp. 1–62. Glasgow: Blackie.

Coutts, R. S. P. 2005. A review of Australian research into natural fiber cement composites. *Cement Concrete Composites* 27:518–526.

Demuth, H., and Beale, M. 2000. *Neural network toolbox user's guide for use with MATLAB Version 4.0.* Nantick, MA: The Math Works, Inc.

European Committee for Standardization. 1994. Fiber-cement profiled sheets and fittings for roofing. Product specification and test methods. EN494, 47 pp.

Guzmán de Villoria, R., and Miravete, A. 2007. Mechanical model to evaluate the effect of the dispersion in nanocomposites. *Acta Materialia* 55:3025–3031.

Haykin, S. 1998. *Neural networks: a comprehensive foundation.* New York: Prentice Hall.

Huber, N., and Tsakmakis, C. 2001. A neural network tool for identifying the material parameters of a finite deformation viscoplasticity model with static recovery. *Computer Methods in Applied Mechanics Engineering* 191:353–384.

Juárez, C., Durán, A., Valdez, P., and Fajardo, G. 2007. Performance of "Agave lech-eguilla" natural fiber in Portland cement composites exposed to severe environment conditions. *Building and Environment* 42:1151–1157.

Kalman, B., and Kwasny, S. C. 1997. High performance training of feedforward and simple recurrent networks. *Neurocomputing* 14:63–83.

Kim, P. J., Wu, H. C., Lin, Z., Li, V. C., de Lhoneux, B., and Akers, S. A. S. 1999. Micromechanics-based durability study of cellulose cement in flexure. *Cement Concrete Research* 29:201–208.

Maggio, R., Franchini, M., Guerrini, G., Poli, S., and Migliaressi, C. 1997. Fiber-matrix adhesion in fiber reinforced CAC-MDF composites. *Cement Concrete Composites* 19:139–147.

Moller, M. F. 1993. A scaled conjugate gradient algorithm for fast supervised learning. *Neural Networks* 6:525–533.

Rillem. Testing methods for fiber reinforced cement-based composites. Mater Constr 1984; 17(102):441–56 (Rilem Draft Recommendations, Technical Committee 49 TFR).

Savastano, H., Jr., Warden, P. G., and Coutts, R. S. P. 2001. Ground iron blast furnace slag as a matrix for cellulose-cement materials. *Cement Concrete Composites* 23:389–397.

Singh, S., Shukla, A., and Brown, R. 2004. Pullout behavior of polypropylene fibers from cementitious matrix. *Cement Concrete Research* 34:1919–1925.

Zhang, J., and Li, V. C. 2004. Simulation of crack propagation in fiber-reinforced concrete by fracture mechanics. *Cement Concrete Research* 34:333–339.

Zhang, Z., Barkoula, N. M., Karger-Kocsis, J., and Friedrich, K. 2003. Artificial neural network predictions on erosive wear of polymers. *Wear* 255:708–713.

Zhang, Z., Klein, P., and Friedrich, K. 2002. Dynamic mechanical properties of PTFE based short carbon fiber reinforced composites: experiment and artificial neural network prediction. *Composites Science and Technology* 62:1001–1009.

Zollo, R. F. 1997. Fiber-reinforced concrete: an overview after 30 years of development. *Cement Concrete Composites* 19:107–122.

8

Fatigue Life Prediction of Fiber-Reinforced
Composites Using Artificial Neural Networks

H. El Kadi and Y. Al-Assaf

CONTENTS

ABSTRACT   This chapter addresses the use of artificial neural networks (ANNs) in the prediction of fatigue life of composite materials. The justification of using soft computing methods including ANNs over other existing methods is initially addressed. Section 8.2 focuses on the various ANN paradigms used in modeling the fatigue life prediction of laminae and laminates. The effect of various network structures and characteristics such as the input parameters, the number of hidden layers, and the number of neurons per hidden layer is discussed. This is followed by a comparison of the different predictions obtained using the aforementioned network architectures and to results obtained using other fatigue life prediction methods. Some ANN design and selection guidelines to follow when predicting the fatigue life of composites using ANNs are also included. The polynomial classifiers, an alternative technique that may address some of the ANN shortcomings, are then introduced. The results obtained using polynomial classifiers are compared to those obtained using the various ANN architectures previously introduced. The chapter concludes by giving a status report about the effectiveness of ANNs in predicting the fatigue life of composites, highlighting the future challenges facing their use.

8.1 Introduction to Fatigue Life Prediction of Composite Materials

The problem of fatigue failure prediction for fiber-reinforced composites is of importance in the design and inspection of composite structures/components used in various engineering applications. In the absence of a well-defined failure criterion that can be used to predict fatigue failure, extensive tests must be carried out for different lay-up configurations and loading conditions. The issue of fatigue life prediction of fiber-reinforced composite materials has been investigated from a number of viewpoints. Proposed methodologies have either been based on damage modeling or on some type of mathematical relationship.

A significant number of research work done in fatigue life prediction of composites was based on the generalization of fatigue failure criteria in metals. In some cases (Hashin and Rotem, 1973), this generalization took into consideration the distinct experimentally observed failure modes (fiber and matrix) and determined a critical fiber orientation angle (θ_c) at which a transition in the failure mode can be observed. These criteria also took into consideration the difference between the composite failure stresses in tension and compression.

The tensor polynomial failure criterion (based on the total strain energy failure theory) that, in its quadratic formulation, can be presented as:

$$F_1\sigma_1 + F_2\sigma_2 + F_6\sigma_{16} + F_{11}\sigma_1^2 + F_{22}\sigma_2^2 + F_{66}\sigma_6^2 + 2F_{12}\sigma_1\sigma_2 = 1 \qquad (8.1)$$

was also used to model the failure of a lamina in plane stress. In this equation, the parameters F_1, F_2, F_6, F_{11}, F_{22}, and F_{66} are strength parameters that are functions on material strengths in the fiber direction, transverse to the fiber direction, and in shear. Parameter F_{12} can be determined experimentally by knowing a biaxial stress state at which the lamina fails. This criterion distinguishes between the tensile and compressive strength of the lamina, and was successfully used in the prediction of the static failure stress of off-axis composite laminae (Awerbuch and Hahn, 1981). Modifying the tensor polynomial criterion by incorporating fatigue functions whereby each strength parameter is regarded as a function of the number of cycles, N, was shown to give reasonably good fatigue life predictions for a lamina under axial loading condition (Tennyson, 1986; Tennyson et al., 1986).

A fatigue damage parameter based on the Smith-Watson-Topper parameter used in metal fatigue was successfully used to predict the fatigue life of off-axis unidirectional fiber-reinforced composites (Plumtree and Cheng, 1999). For small plastic strains, the Smith-Watson-Topper parameter allowed for comparison of different mean normal stresses by converting them to an equivalent zero mean stress range for a particular life. The Smith-Watson-Topper parameter written as:

$$P_N = \sigma_{max} \Delta \varepsilon \tag{8.2}$$

where σ_{max} is the maximum applied cyclic stress, $\Delta \varepsilon$ is the tensile strain range, and P_N is the mean normal stress fatigue parameter. This parameter, which combines both normal and shear contributions, is expressed in the same units as strain energy density and, for a unidirectional off-axis composite, has the following form (Petermann and Plumtree, 2001):

$$\Delta W^* = \frac{1-R^2}{2}(\sigma_{22}^{max}\varepsilon_{22}^{max} + \tau_{12}^{max}\gamma_{12}^{max}) \tag{8.3}$$

where σ_{22}^{max} and ε_{22}^{max} are the components of the normal in-plane stress and strain perpendicular to the fiber direction, and τ_{12}^{max} and $\gamma_{12}^{max}/2$ are the components of the in-plane maximum shear stress and strain, respectively, acting on the fiber-matrix interface. The predicted results were found to be in good agreement with experiments for different fiber/load angles and stress ratios.

Based on the premise that the damage caused in the material due to cyclic loading is related to the mechanical energy input, the strain energy criteria used in metals was modified to be used as a fatigue failure criterion for fiber-reinforced laminae (Ellyin and El Kadi, 1990). The power-law relation used for metals was modified to include the fiber orientation angle:

$$\Delta W(\theta) = \kappa(\theta)N_f^{\alpha(\theta)} \tag{8.4}$$

A normalized form of the criterion was shown to collapse all S-N curves obtained for various fiber orientations onto a single curve. This criterion was later extended (El Kadi and Ellyin, 1994) to account for the effect of the stress ratio ($R = \sigma_{min}/\sigma_{max}$). In this case, all data points obtained from different combinations of fiber orientation angles and stress ratios can be collapsed on one single curve:

$$\Psi = \Psi^+ + \Psi^- = \frac{\Delta W^+}{W_f^+} + \frac{\Delta W^-}{W_f^-} \tag{8.5}$$

where Ψ is the normalized form of the strain energy, ΔW^+ and ΔW^- are the areas under the stress/strain diagram associated with the tensile and compressive parts, respectively, and W_f^+ and W_f^- are the monotonic values of the strain energy under tensile and compressive loads, respectively.

A different strain energy failure criterion (Varvani-Farahani et al., 2007) based on the mechanism of fatigue cracking successfully correlated the fatigue lives of unidirectional composites at various off-axis angles and stress ratios. The fatigue cracking was determined in the three damage regions of matrix (I), fiber (II), and fiber-matrix interface (III) as the number of cycles progresses. The total fatigue damage in unidirectional off-axis composites is calculated by integrating the fatigue damage within each of the three regions, that is,

$$\Delta W = \Delta W_{\mathrm{I}} + \Delta W_{\mathrm{II}} + \Delta W_{\mathrm{III}} \tag{8.6}$$

Results of fatigue damage assessment revealed that this damage parameter correlated the fatigue data of several composite materials with a high degree of success.

Incorporating a micromechanics-based evaluation of damage evolution with a semiempirical fatigue failure criterion, a progressive damage parameter (Akshantala and Talreja, 2000) was tested for cross-ply laminates under cyclic tension. In this instance, not only the fatigue lives of laminate but also the damage sustained at any intermediate number of cycles could be predicted. This is especially attractive for structural applications because inspection intervals can be decided based on the model predictions and the number of inspections can also be significantly reduced. The change in crack density in a laminate loaded in fatigue at any given number of cycles can be written as:

$$\eta = \eta(\sigma_{\mathrm{max}}, l, M, K) \tag{8.7}$$

where l is the delamination length, M denotes the material properties of the laminate, and K denotes the geometry of the laminate, for example, thickness of the cracking plies. A power-law relation can be assumed for crack density change with cycles. The predicted results are compared with experimental data for several glass/epoxy and carbon/epoxy laminates, and a good agreement is found.

Based on the hypothesis that accurate fatigue strength prediction depends on the allowable fatigue stresses of its constituent fiber and matrix materials, another micromechanics model (the bridging model; Huang, 2001) was successfully used to estimate the fatigue failure strength and S-N relationships of multidirectional composite laminates. This model was incorporated with a fatigue failure criterion similar to the maximum normal stress criterion of isotropic materials subjected to arbitrary cyclic loading. Only the constituent fatigue properties and the fiber volume fraction are necessary, reducing the number of required input data. When monolithic material tests are not available, these parameters can be retrieved using measured fatigue data of a lamina ply in two different off-axial directions or using those of two different angle-ply laminates. The predictions for a number of angle ply glass/epoxy laminates under tensile fatigue showed a good correlation between the predicted results and available experimental data.

Incorporating the effect of stress ratio in a fatigue failure criterion of composites is imperative to the accurate prediction of their fatigue life. In addition to those already mentioned (Plumtree and Cheng, 1999; Petermann and Plumtree, 2001; El Kadi and Ellyin, 1994; Varvani-Farahani et al., 2007), the effect of nonnegative mean stress on the off-axis fatigue behavior of unidirectional composites was addressed by Kawai (2004) and Kawai and Suda (2004) through the development of a nondimensional effective stress that considers the effects of stress ratio and the state of stress:

$$\Sigma^* = \frac{\frac{1}{2}(1-R)\sigma_{max}^*}{1-\frac{1}{2}(1+R)\sigma_{max}^*} \tag{8.8}$$

where σ_{max}^* is the maximum nondimensional effective stress. A master S-N relationship can be obtained by plotting the effective stress against the number of reversals to failure. This fatigue model was shown to adequately describe the stress ratio dependence as well as the fiber orientation dependence of the off-axis fatigue behavior under nonnegative mean stresses.

To include the effect of load frequency to that of the stress ratios, a fatigue model using a well-defined minimum number of tests was used to predict the fatigue behavior of a composite material (Epaarachchi and Clausen, 2003). The model that includes the nonlinear effect of stress ratio and load frequency on the fatigue life can be mathematically represented as:

$$\sigma_u - \sigma_{max} = \alpha\, F(R, \sigma_{max}, \sigma_u)\frac{1}{f^\beta}(N^\beta - 1) \tag{8.9}$$

where f is the load frequency, and α and β are material parameters. The model showed excellent agreement between predictions and experimental data for a wide range of glass fiber-reinforced plastics. Furthermore, it was shown that using a set of standard fatigue tests, preferably $R = 0$ or 0.1 for tension-tension loading and $R = 10$ for compression-compression loading, over a suitable range of frequency to avoid heating in the specimen, are enough to predict the whole fatigue life (S-N) curves for a particular glass fiber-reinforced plastic composite material.

Lamina-to-laminate approaches to predict the fatigue behavior of composite laminates have the merit of theoretically being applicable to a laminate with any lay-up while requiring less experimental input. However, since constrained laminae behave differently from unconstrained ones, obtained results are usually inaccurate. A fatigue criterion taking into account the

constrained condition of the individual laminae in a multidirectional laminate would render the lamina-to-laminate approach attractive. For constant
amplitude cases, a lamina failure model is given by (Fawaz and Ellyin,
1995):

$$S(a_1, a_2, N) = f(a_1, a_2)[\gamma m \log(N) + b] \tag{8.10}$$

where S is the cyclic stress, m is a parameter obtained from a reference S-N
curve for a constrained lamina, a_1 and a_2 are the biaxial ratios, γ is a parameter
depending on the off-axis angle of the adjacent layers, and b is the stress level
at which stress concentrations start developing. This theory was shown to
predict the fatigue failure results of a large family of laminates while requiring a minimal amount of experimental input.

In many instances, one would like to have some initial information on the
behavior of a certain composite material even if the information is not very
accurate. For example, as a new material is being developed, one would like
to have an acceptable prediction as to the expected fatigue properties of this
new material for different lay-up configurations, loading conditions, as well
as environmental conditions. Super Mic-Mac, an Excel-based program, was
used for such purpose (Sihn and Tsai, 2005). Super Mic-Mac is not intended
to replace more exact analyses such as the finite element analysis; it, however, helps to answer such questions as "what if" early on in the process.

From the above-mentioned work as well as other research published on the
fatigue life prediction of composite materials, it can be concluded that modeling the cyclic behavior of composites can be successfully and accurately
done by many existing methods. One of the problems involves the numerous parameters developed to better predict the fatigue life. Another problem
relates to the large number of tests of S-N curves required to ensure good
representation of fatigue failure. This is sometimes impossible to obtain at
the preliminary design stages. It is therefore necessary to introduce some
method that allows for some type of acceptable fatigue life prediction at that
early stage without the need for costly, time-consuming experiments. This
method should be based on appropriate but uncomplicated parameters associated with the material used and the loading it is subjected to. The use of
artificial neural networks (ANNs) to predict the fatigue of composites could
be an excellent method to obtain such results at affordable cost.

8.2   Artificial Neural Networks

The biologically motivated (Kartalopoulos, 1996) computing paradigm of
ANNs has emerged as a superior modeling tool to deal with situations when

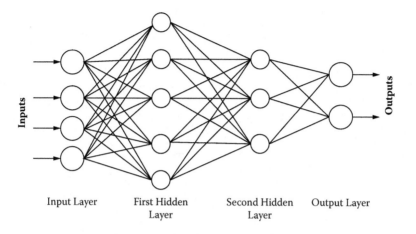

FIGURE 8.1
General configuration of a feedforward artificial neural network (El Kadi, 2006, with permission).

it is difficult to find an accurate mathematical-based solution and the available data are incomplete, noisy, or complex. ANNs generally consist of a number of layers (Figure 8.1); the layer where the input patterns are applied is called the *input layer*. This layer could include the static and cyclic properties of the composite material under consideration, its lay-up, the applied stress, the stress ratio, etc. The layer where the output is obtained is called the *output layer*, which could, for example, contain the fatigue life of this composite under the specific loading conditions. In addition, there may be one or more layers between the input and output layers called the *hidden layers*, which are so named because their outputs are not directly observable (Skapura, 1996). The addition of hidden layers enables the network to extract higher-order statistics, which are particularly valuable when the size of the input layer is large (Schalkoff, 1997). Neurons in each layer are fully or partially interconnected to preceding and subsequent layer neurons with each interconnection having an associated connection strength (or weight). The input signal propagates through the network in a forward direction, on a layer-by-layer basis. These networks are commonly referred to as multilayer feedforward neural networks (FNNs). Many reports discuss the development and theory of ANNs (see, e.g., Kartalopoulos, 1996; Skapura, 1996; Schalkoff, 1997; Jang et al., 1997; Haykin, 1998; Aleksander and Morton, 1990; Fausett, 1994; Bishop, 1995; Swingler, 1996; McCulloch and Pitts, 1943; Rosenblatt, 1958, 1962).

The back-propagation training algorithm (Kartalopoulos, 1996; Skapura, 1996) is commonly used to iteratively minimize the following cost function with respect to the interconnection weights and neuron thresholds:

$$E = \frac{1}{2} \sum_{1}^{P} \sum_{i=1}^{N} (d_i - O_i)^2 \qquad (8.11)$$

where P is the number of experimental data pairs used in training the network and N is the number of output parameters expected from the ANN. d_i and O_i could be the experimental number of cycles to failure and the current life prediction of the ANN for each loading condition i, respectively. Iteratively, the interconnection weights between the jth node and the ith node are updated as:

$$w_{ji}(t+1) = \alpha w_{ji}(t) + \eta x_i f'(net_j^k) \sum_{i=1}^{N} \left(d_l - O_l\right) f'(net_i^o) w_{lj} \qquad (8.12)$$

where α is a momentum constant, η is the learning rate, x_i is the input pattern at the iterative sample t, net_N^o is the input to node N at the output layer, and net_j^k is the input to a node j in the kth layer and the function f' is the derivative of the neuron activation function. The learning rate determines what amount of the calculated error sensitivity to weight change will be used for the weight correction. It affects the convergence speed and the stability of weights during learning. The "best" value of the learning rate depends on the characteristics of the error surface. For rapidly changing surfaces, a smaller rate is desirable whereas for smooth surfaces, a larger value of the learning rate will speed up convergence. The momentum constant (usually between 0.1 and 1) smoothes weight updating and prevents oscillations in the system and helps the system escape local minima in the training process by making the system less sensitive to local changes. As much as the learning rate, the momentum constant "best" value is also peculiar to specific error surface contours. The training process is terminated either when the mean square error (MSE), root mean square error (RMSE), or normalized mean square error (NMSE), between the actual experimental result and the ANN prediction obtained for all elements in the training set, has reached a prespecified threshold or after the completion of a prespecified number of learning epochs.

Although all neural network models share common operational features, their input requirements and modeling and generalization abilities are different. Consequently, each paradigm would have advantages and disadvantages depending on the particular application, and selecting the appropriate network class with suitable parameters is vital to ensure a successful application. More details about the various ANN structures, their similarities, and their differences can be found in the reports of Kartalopoulus (1996), Skapura (1996), Schalkoff (1997), Jang et al. (1997), Haykin (1998), Aleksander and Morton (1990), Fausett (1994), Bishop (1995), Swingler (1996), McCulloch and Pitts (1943), Rosenblatt (1958, 1962), Kohonen (1984), and Powell (1985). The following neural network classes have been used in the modeling and simulation of fiber-reinforced composite materials.

Modular neural networks (MNNs). The central idea behind such networks is task decomposition (Skapura, 1996; Schalkoff, 1997; Jang et al., 1997; Haykin, 1998). Using an MNN, the task is split up among several local neural networks (subnetworks) not communicating with each other. The most attractive feature of MNN architecture is that each subnetwork may model one of the fatigue parameters as an input (R, θ, or σmax). The generalization requirement for each subnetwork is reduced in comparison with a single ANN that must learn the entire task. A gating network determines the contribution of each subnetwork to the total output as well as the range of the input space each network should learn. The main problem with modular networks is how to choose modules and how to structure the problem. In simple cases, human design knowledge may be used to restructure the problem and the data set for the modular solution. This approach is only applicable for a very limited domain of problems. Optimization techniques could be used to find optimal network architecture for complicated applications such as fatigue life predictions; however, an adequate set of data is needed to generate acceptable results.

Self-organizing feature maps (SOFMs). The SOFM is a clustering algorithm (Kohonen, 1984) that creates a map of relationships among input patterns. The map is a reduced representation of the original data. It has a single layer of nodes and uses a distance metric to find the output node closest to a given input pattern. The SOFM output nodes do not correspond to known classes but to unknown clusters that the network autonomously finds in the data. During training, the SOFM finds the output node that has the least distance from the training pattern. It then changes the node's weights to increase the similarity to the training pattern, and it influences the weights of the neighboring nodes even though they have only random relationships to the training pattern. Different patterns trigger different winners that influence different neighbors. The overall effect is to move the output nodes to positions that map the distribution of the training patterns. After training, each node's weights model the features that characterize a cluster in the data. As in the case with MNN, having enough and experimentally sound data would enhance the accuracy of the SOFM in fine clustering the input-output space.

Radial basis function (RBF) networks. The RBF is a classification and functional approximation paradigm developed by Powell (1985). These networks are nonlinear hybrid networks usually containing a single hidden layer. The centers and widths of the activation functions are obtained by unsupervised learning, whereas supervised learning is used to update the connection weights between the hidden and output layers. It is claimed that these networks learn faster than the FNN and need a smaller number of training data; their generalization capabilities, however, are limited (Skapura, 1996; Jang et al., 1997). As previously mentioned, neural networks were used in fatigue prediction as an alternative to save time and avoid costly experimentations, and having limited generalization with RBF may indicate that they are not the type of networks to be considered for such applications.

Principal component analysis (PCA) networks. The PCA involves a mathematical procedure that transforms a number of (possibly) correlated variables into a (smaller) number of uncorrelated variables called principal components. The first principal component accounts for as much of the variability in the data as possible, and each succeeding component account for as much of the remaining variability as possible. After the PCA components are obtained, the supervised segment of the network performs the nonlinear modeling using the classical feedforward method. Another way to view PCA is determining which input has the greatest influence on the output.

Recurrent neural networks (RNN). An RNN distinguishes itself from a FNN in that it has at least one feedback loop. For example, a RNN may consist of a single layer of neurons with each neuron feeding its output signal back to the inputs of all the other neurons. The presence of feedback loops has a profound impact on the learning capabilities of the network and its performance. Moreover, the feedback loops involve the use of particular branches composed of unit-delay elements resulting in a nonlinear behavior assuming that the neural network contains nonlinear units (Haykin, 1998). Since fatigue predictions are not a temporal process, one would not see the advantage of using RNNs to predict fatigue life. However, benefits may result because of the feedback structure pattern of the network, such that the overall weight calculations may result in obtaining a more optimal static neural network.

The FNNs and the MNNs have been found by many research groups (El Kadi, 2006; Zhang and Friedrich, 2003; Hajela, 2002) as the most adequate neural network architectures to use in predicting the mechanical behavior of composite materials.

8.3 Fatigue Modeling of Fiber-Reinforced Polymeric Composites Using ANNs

8.3.1 Fatigue of Laminae

Stress-life experimental data for glass fiber/epoxy laminae with five fiber orientation angles ($\theta = 0°$, $19°$, $45°$, $71°$, and $90°$) subjected to three values of the stress ratio ($R = -1$, 0, and 0.5) were used by Al-Assaf and El Kadi (2001) to train and test the ANN. A variation of input parameter combinations was attempted to minimize the RMSE obtained while maximizing the corresponding correlation coefficient. These included the fiber orientation angle, the minimum applied stress, the maximum applied stress, the stress ratio, the

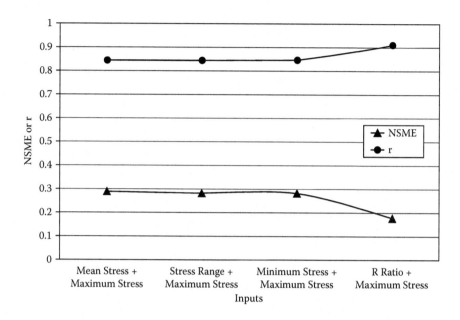

FIGURE 8.2
Results of trials to obtain the optimum inputs to the ANN (Al-Assaf and El Kadi, 2001, with permission).

stress range, and the mean stress. The best predictions were obtained using the fiber orientation angle, the stress ratio, and the maximum stress as inputs, with the output being the number of cycles to failure (Figure 8.2). Since the range of number of cycles to failure varied between 300 and 1,000,000, training the network to learn over such a wide range will produce unacceptable and unbalanced modeling performance. This will occur since the ANN will strive to minimize the overall MSE for all input patterns. Hence, minimizing the difference between the network output and observed data for the high number of cycles would be at the expense of the lower values. Classical normalization, where the range is scaled between 0 and 1, will also not solve the problem since smaller values of life cycles will be very close to zero. To make the output amenable for successful learning, the logarithmic values for the number of cycles to failure were considered; this reduced the scale to lie between 0.5 and 6. The maximum stress applied varied between 13 and 760 MPa. The fiber orientation angle had five values: 0°, 19°, 45°, 71°, and 90°. The values for σ_{max} and θ were normalized between 0 and 1 for network training and testing. R had three values: 0, 1, and –1. The effect of the number of hidden layers as well as the number of neurons per layer was also investigated. It was found that a feedforward ANN with one hidden layer containing 12 neurons was the most appropriate considering the accuracy of the predictions obtained as well as the complexity of the network and the extent of the required calculations. This follows the typical rule of thumb stating that

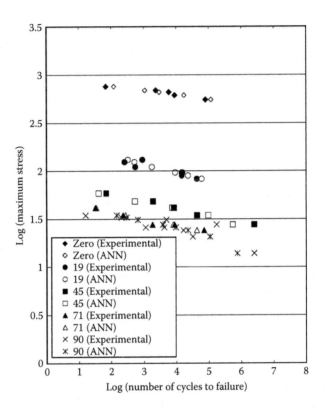

FIGURE 8.3
Comparison between experimental data and ANN predicted number of cycles to failure for *R* = −1 (Al-Assaf and El Kadi, 2001, with permission).

the number of neurons in the hidden layer should be 3–4 times the number of input parameters to the ANN. Although a relatively small number of data points (92 experiments) was used to train the network, the predictions obtained using the Neurosolution software (NeuroSolutions) were found to be comparable to experimental data (Figure 8.3).

To investigate the effect of various neural network architectures on the fatigue life predictions of composite laminae, the results previously obtained using FNNs were compared to those predicted using MNN, RBF, SOFM, and PCA (El Kadi and Al-Assaf, 2002b). It was concluded that the modular networks, with their ability to decompose the modeling task, showed the most significant improvement as its use resulted in a reduction of NMSE (from 14.2% to 5.7%) and an increase in the correlation coefficient, *r* (from 92.6% to 97.3%; Table 8.1). Figure 8.4 also shows the comparison between the predicted results using FNN and MNN for a fiber orientation angle of 45° for the various stress ratios used.

One of the problems faced when using ANNs is that of overcorrelation due to the network, not only learning the trend of the experimental data,

TABLE 8.1

Comparison of NMSE and *r* for Various Neural
Network Structures (El Kadi and Al-Assaf, 2002b)

Network	NMSE	*r*
FNN	0.14273	0.9260
MNN	0.05668	0.9726
RBF	0.6600	0.4703
SOFM	0.0743	0.9624
PCA	0.1402	0.9280

NMSE, normalized mean square error; FNN, feedforward neural network; MNN, modular neural network; RBF, radial basis function; SOFM, self-organizing feature map; PCA, principal components analysis.

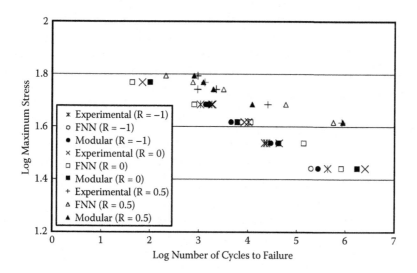

FIGURE 8.4
Comparison between fatigue life prediction using FNN and MNN for θ = 45° (El Kadi and Al-Assaf, 2002b, with permission).

but also the existing scatter in the data. The same issue labeled as overfitting was discussed by Zhang and Friedrich (2003), and defined as the network having memorized the training example, but not having learned to generalize to new situations. A suggested solution to this problem could be through the use of a third group of data set called the validation set (aside from the training and testing data sets). If the validation error increases for a specific number of iterations due to overfitting, the training is stopped early and the weights and biases are returned to the minimum of the validation error. In an attempt to estimate the effect of overcorrelation, the strain energy density defined as:

$$\Delta W = \frac{1}{2}\left(\sigma_{max}\varepsilon_{max} - \sigma_{min}\varepsilon_{min}\right) \tag{8.13}$$

was used as an input to the ANN (El Kadi and Al-Assaf, 2002a). Using the energy rather than the maximum stress, the stress ratio and the fiber orientation as inputs also reduced the complexity of the ANN. Although the resulting correlation obtained was lower (79.3%) and the NMSE was higher (37%) than those obtained when the three parameters were used as inputs (Al-Assaf and El Kadi, 2001; El Kadi and Al-Assaf, 2002b), the overall predictions seemed more realistic.

Although fatigue predictions are not a temporal process, using RNNs to predict fatigue life may improve those predictions due to having a feedback-structure pattern that might lead to a more optimal static neural network. As shown in Figure 8.5, RNNs are multilayer back-propagation networks, with the addition of a feedback connection from the output of the hidden layer to its input. This feedback path allows RNNs to learn to recognize and generate temporal patterns as well as spatial patterns (Elman, 1990).

RNN differs from conventional FNNs in that the hidden layers have a recurrent connection. The delay in this connection stores values from the previous time step, which can be used in the current time step. Using the RNN, the fatigue behavior of unidirectional glass fiber/epoxy composite laminae under tension-tension and tension-compression loading was predicted. The MATLAB software and its toolboxes were used in training and testing the data. Figure 8.6 shows that the results obtained using RNNs are superior to those obtained using the traditional FNNs (Al-Assaf and El Kadi, 2007).

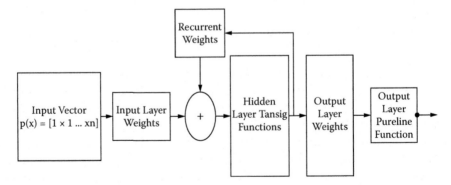

FIGURE 8.5
A two-layer recurrent neural network architecture (Al-Assaf and El Kadi, 2007, with permission).

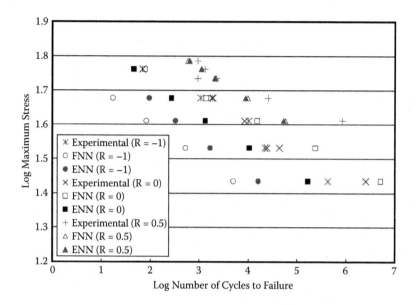

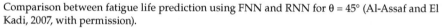

FIGURE 8.6

Comparison between fatigue life prediction using FNN and RNN for θ = 45° (Al-Assaf and El Kadi, 2007, with permission).

8.3.2 Fatigue of Laminates

In an attempt to use ANNs for fatigue life prediction of composites, a back-propagation ANN was used to model the cyclic stress-strain behavior of [±θ] graphite-epoxy laminates (Pidaparti and Palakal, 1993). The input to the neural network included the fiber angle, cycle number, index (1 or 0) for loading and unloading, stress, and the output from the network is the total strain. A neural network with a <4-17-17-1> configuration was submitted to a training data set consisting of 180 data points using the same learning coefficient and momentum term values the authors used for the monotonic case. Normalized experimental data for θ = 0°, 30°, 50° orientation were used in training the neural network. The network converged at approximately 10,000 iterations. The study shows that a neural network can be effectively used to represent the cyclic stress-strain behavior in composites (Figure 8.7). With no other experimental data available in the literature, the authors did not make any other predictions for the cyclic stress-strain behavior.

A more detailed study (Lee et al., 1999) involving data obtained for [(±45,0₂)₂]ₛ laminates under T-T, T-C, and C-C loading cases was conducted to evaluate the performance of ANNs in fatigue life prediction of composites. The software used was a commercial neural network development tool, NeuDESK (NEUDESK). The neural network was trained using data from more than 400 fatigue tests on carbon-epoxy materials over a range of five R ratios, from +0.1 to +10. Experiments showed that the RMSE changes with the number of nodes in the hidden layer.

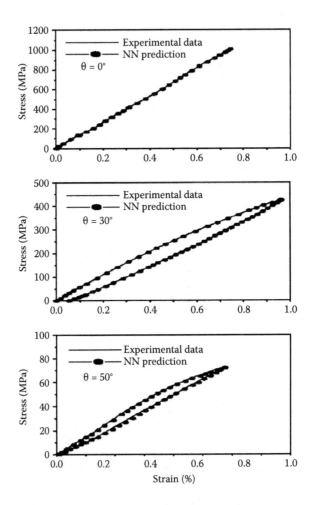

FIGURE 8.7
Comparison of neural network predictions with experimental data for graphite/epoxy [±θ] laminate under cyclic loading (Pidaparti and Palakal, 1993, with permission).

The optimum architecture was found to contain a ratio of three hidden-layer nodes to every input node. This conclusion was also made when using ANNs in many noncomposite materials applications. As in the work of Al-Assaf and El Kadi (2001), a variety of input parameter combinations was attempted to minimize the RMSE obtained. The best results were obtained using the maximum and minimum stress and failure probability (Figure 8.8). It should be noted that these input parameters are slightly different than those found to produce the optimum results (see Al-Assaf and El Kadi, 2001). The output in all cases was the number of cycles to failure. The number of stress ratios used for training and its effect on the accuracy obtained was also studied; although increasing the number of stress ratios improved the accuracy, the rate of this increase fell significantly when the number of R ratios increased beyond three. The chosen R ratios, however, should contain examples of the three modes of cycling: all tension, all compression, and mixed tension/compression. The authors concluded that ANNs can be trained to model constant-stress fatigue behavior at least as well as other prediction methods and can provide accurate results from quite small experimental databases.

To take into consideration the orientation of the various laminae in angle-ply laminates, two or three input parameters (to the neural network) were used (Aymerich and Serra, 1998): the first parameter was associated with the number of cycles to failure, whereas the remaining parameters identified the lay-up of the laminate; the output was the fatigue strength of the laminate. To reduce the number of input parameters, laminate geometry was identified by the set of orientation angles of each couple of balanced layers: a single parameter was needed to identify the $[\pm 35]_{4s}$ laminate, but the set (22.5, 67.5) was needed to identify the $[-22.5/22.5/-67.5/67.5]_{2s}$ laminate. A back-propagation ANN was used to predict the fatigue strength of APC2, an AS4 graphite in a thermoplastic peek matrix with an average fiber content of 63%. Angle-ply laminates $[\pm\theta]_{4s}$ (with θ = 0°, 15°, 25°, 35°, and 45°) as well as $[0/90]_{4s}$ and $[-22.5/22.5/-67.5/67.5]_{2s}$ were tested at a stress ratio of 0.1. The sequences examined assured a wide range of failure modes ranging from matrix-dominated to fiber-dominated behavior. Although this approach took into consideration the orientation of the various laminae in the laminate, it failed to account for their staking sequence, which is an important factor in the failure of laminates. For the learning stage of the ANN, various strategies were used. In the first (case 1), all but one of the angle-ply laminates were used for training; the network was required to predict the S-N behavior of the removed sequence. The predictions obtained for the $[\pm 15]_{4s}$, $[\pm 25]_{4s}$,

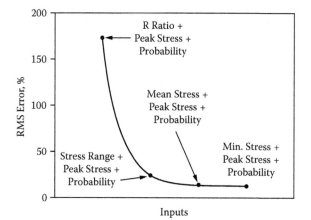

FIGURE 8.8
Results of trials to establish the optimum inputs to the ANN (Lee et al., 1999, with permission).

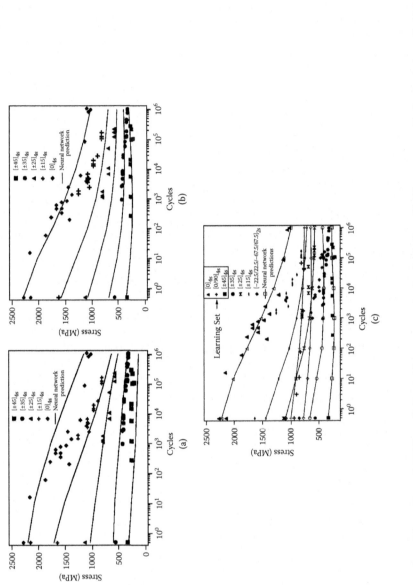

FIGURE 8.9

Comparison between experimental results and ANN predictions: (a) one-parameter laminate (case 1); (b) one-parameter laminate (case 2); (c) two-parameter laminate (Aymerich and Serra, 1998, with permission).

and $[\pm35]_{4s}$ laminates were in good agreement with the experimental results, and those for the $[0]_{4s}$ and $[\pm45]_{4s}$ laminates were not (Figure 8.9a). This is explained by the fact that in these two cases, the neural network was asked to extrapolate results beyond the training range. A second strategy (case 2) was to use only the results from the two extreme cases: $[0]_{8s}$ and $[\pm45]_{4s}$ as inputs to the network. The predictions obtained were acceptable for the $[\pm30]_{4s}$ laminates, but were not as good for the $[\pm15]_{4s}$ and $[\pm25]_{4s}$ laminates (Figure 8.9b). This was attributed to the sudden change in failure mode between the $[0]_{8s}$ and the $[\pm15]_{4s}$ laminates. Better results would have definitely been obtained if the data set contained a laminate with fiber orientation between the $0°$ and the $15°$ fiber angles. A data set using the angle-ply laminates as well as the $[0/90]_{4s}$ and $[-22.5/22.5/-67.5/67.5]_{2s}$ laminate resulted in unsatisfactory predictions where, in some cases, errors of up to 50% were obtained (Figure 8.9c). This work showed that, although the use of ANN resulted in some good fatigue life prediction in fiber-reinforced composites, a large set of experimental data representative of the failure modes and the laminate sequences is necessary.

To obtain a faster conversion and a higher efficiency, various training algorithms have been used in predicting the fatigue behavior of composites. An ANN-based model with a modified back-propagation feedforward technique using the steepest descent algorithm was used to describe the split growth in notched AS4/3501-6 graphite epoxy quasi-isotropic $[0/45/-45/90]_{3s}$ laminate under tension-dominated fatigue with a constant stress ratio of 0.1 (Choi et al., 2003). The neural network used was structured with $\log_{10}(N)$ and σ_{max} as the input vector components and the split length as the output. The neural network package available in MATLAB (MATLAB) was used in the analysis. Although ANNs with two hidden layers are only used when the system is very complex and the results obtained using one hidden layer are unsatisfactory, the three models used in this study had two hidden layers. These networks were trained using the constant amplitude fatigue data and gave better predictions compared to a power-law method often used in fatigue analysis (Figure 8.10). A linear cumulative damage growth rule was used in conjunction with the power-law method and the ANN model to predict the split growth under spectrum fatigue. In this case, since the stress ratio varied from one stress level to the next, the value of the stress ratio was disregarded in the predictions and only the maximum fatigue stress was used as an input to the neural network. Once again, the ANN model resulted in good predictions; this was attributed to the network's ability to capture more of the nonlinear characteristics compared to the simple power law.

Constant fatigue life diagrams are typically used to show the overall behavior of a material under cyclic conditions for various stress ratios. Building such a diagram requires a large number of experimental results and the successful use of ANNs in building constant fatigue life diagrams for composite materials would reduce the number of experiments needed. Twelve different S-N curves obtained from the literature for a plastic reinforced with fiberglass

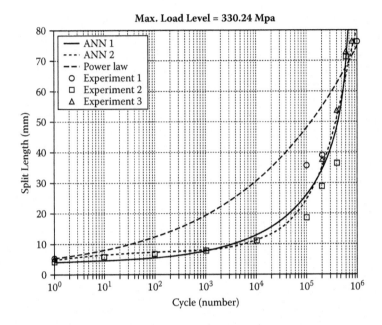

FIGURE 8.10
Split growth at 80% Notched tensile strength (NTS)—comparison between experimental results, power law, and ANN predictions (Choi et al., 2003, with permission).

denoted as DD16 with a $[90/0/\pm45/0]_s$ configuration were used in conjunction with the multilayer FNNs trained with the back-propagation algorithm used (Freire et al., 2005). The input parameters were the mean stress and the number of cycles, whereas the lone output parameter was the alternating stress. One hidden layer with bias and sigmoid activation function containing 2–30 neurons was used in the study. The output neurons used a linear activation function. Training of the ANN was attempted with data obtained from three, four, five, and six stress ratio values. Even when only three S-N curves were used in the training, satisfactory results were obtained (Figure 8.11a). On the other hand, when the number of training S-N curves was increased, a more reliable solution was obtained (Figure 8.11b). Using an MNN in the development of the constant life diagram (Freire et al., 2007) led to better results compared to the traditional FNN. As for the FNN study, the results obtained using six values of the stress ratio during the training produced the best results (Figure 8.12); however, increasing the number of modules gives better predictions only if the training set is also increased.

A back-propagation algorithm based on least square method was used (Mathur et al., 2007) to predict the fatigue behavior of several carbon fiber-reinforced plastics with lay-ups of $[\pm45,0_2)_2]_s$ and $[(\pm45,90,0)_2]_s$. The experimental database used contained results spanning stress ratios from 10 to −3.33. Various static and cyclic material properties were used as inputs to the ANN,

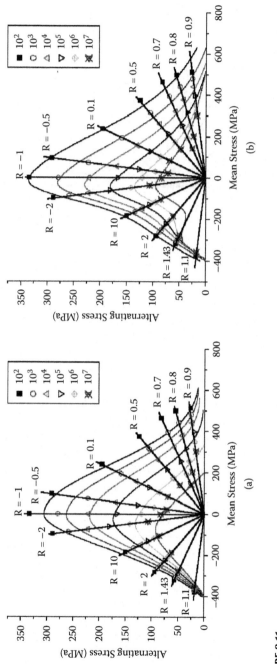

FIGURE 8.11
Constant life diagrams made from FNNs using: (a) a data set of 3R values, (b) a data set of 6R values (Freire et al., 2005, with permission).

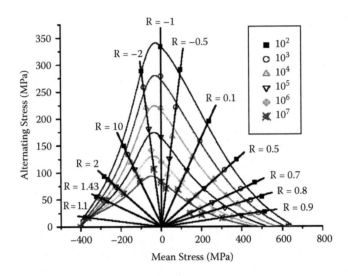

FIGURE 8.12
Constant life diagram built using an MNN with a 6R training set (Freire et al., 2007, with permission).

whereas fatigue life was used as the output parameter. The input parameters used included volume fraction, tensile modulus and strength, compression strength, failure strain, maximum and minimum applied stress, stress ratio, as well as probability of failure and statistical parameters of fatigue life. The architecture of the ANN used included two hidden layers with 18 and 6 neurons in the first and second hidden layers, respectively. The predictions obtained showed that the ANN used resulted in good accuracy as more than 92% of the predictions had a less than 5% error.

A recently published study (Freire et al., 2009) assessed the applicability of two ANN architectures (multilayer feedforward and modular) in the predictions of fatigue life in composites compared to the equation developed by Adam et al. (1989) for modeling the constant life diagram. Glass fiber-reinforced plastics in the form of laminar structures with distinct stacking sequences were used in the study. Reinforcements in the form of mats (M) of short fibers and bidirectional textile fabric (T) were used. Two lay-up sequences were investigated in this work: [M/T/M/T/M]$_s$ and [M/T/M/T/M/M/T/M/T/M/T/M]. These materials were tested for six different stress ratios: $R = 1.43, 10, -1.57, -1, 0.1$, and 0.7. Two training sets were used in training both ANN structures considered: one with three stress ratio values and the other with four stress ratio values. Results showed that modeling of constant life diagram can be done using Adam's equation, but a large number of tests of S-N curves are required to ensure good representation of fatigue failure. This does not occur for the neural network modeling of constant life diagram when excellent results were obtained for a much smaller set of

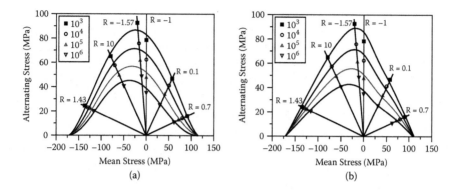

FIGURE 8.13
Constant life diagram (a) modeled from Adam's equation and (b) obtained using an MNN (Freire et al., 2009, with permission).

experimental data. Analysis of the model created with a modular network architecture showed that this network produced much better results than those obtained by both the feedforward network and by Adam's equation for all the laminates analyzed, including a laminate obtained from the literature, in which greater generalization capacity and robustness was achieved (Figure 8.13).

8.3.3 Predicting the Fatigue Life of New Materials

In all previously mentioned studies using ANNs to forecast the fatigue life of fiber-reinforced composites, the authors predicted failure with respect to various design parameters (e.g., fiber orientation, stress ratio) of one specific material. This means that the ANN was trained on a specific material and was required to predict the cyclic behavior of the same material under different conditions. It should be mentioned, however, that one of the anticipated benefits of the successful application of ANNs would be that it would be possible to predict the cyclic behavior of a material for which no fatigue data are available by using the known characteristics of other composites. Some of the published research (Vassilopoulos and Bedi, 2008) concluded that although ANNs have proved to be very accurate in interpolating a material's behavior inside a known database, they cannot extrapolate outside the database nor for different loading conditions or different materials.

Another attempt to predict the cyclic behavior of a composite material not used in training the ANN (Lee et al., 1999) led to unsatisfactory results with an average RMSE in the order of 100%. Data from the four carbon fiber-reinforced plastics were used to predict the fatigue properties of the fifth type of carbon fiber-reinforced plastic. Monotonic mechanical properties (tensile and compressive strengths) were used in the training. It is noteworthy that the level of error obtained was within the normal spread of data for composite materials. On the other hand, using the network trained on data for carbon fiber-reinforced plastics to predict the behavior of a similar structure made of glass/epoxy, resulted in an RMSE in the order of 170%. It was concluded that there seemed to be little prospect of transferring the predictive capability of a network with any degree of accuracy from one family of composites to another.

Better fatigue life predictions were obtained when an MNN was used to predict the number of cycles to failure for different composite materials (El Kadi and AL-Assaf, 2007). Experimental fatigue data obtained for unidirectional fiber-reinforced laminates were used to predict the cyclic behavior of a composite made of another material. Seven different materials were used in this study; data were collected for a variety of published fatigue data with a stress ratio of 0.1. The input parameters to the ANN comprised a combination of monotonic and cyclic properties. The monotonic properties used as input parameters were as follows: modulus of elasticity in the direction of the fiber and perpendicular to the fibers, tensile strength of the lamina in the fiber direction and perpendicular to the fibers, fiber volume fraction, and fiber orientation angle. The fiber volume fraction was later disregarded since its variation (for the considered materials) was minimal and its effect on the prediction negligible. In addition, the maximum applied stress, σ_{max}, was also supplied to the ANN as an input parameter. The sole output from the ANN is the number of cycles to failure (N_f). Contrary to previously published research, preliminary results are encouraging with an RMSE in the order of 38%. The effects of ANN architecture, training algorithm, as well as number of neurons per hidden layer were later considered (Al-Assadi and El Kadi, 2009) to identify the ANN giving the optimum fatigue life prediction. The following training functions were used in predicting fatigue life: resilient back-propagation, Gauss data archives, variable learning rate back-propagation, and gradient descent with momentum. The results showed that ANNs can be accurately used to predict the fatigue failure of a composite material not used in the training of the network. The cascade neural network with 20 neurons gave the best results with 10% RMSE (Figure 8.14).

8.4 Guidelines for Using ANN in Fatigue Life Predictions of Composite Materials

The following list contains some recommended guidelines when using ANNs to predict the fatigue failure of composites:

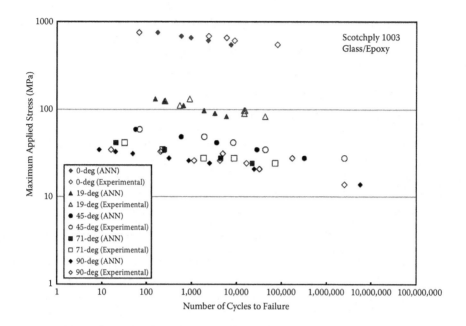

FIGURE 8.14
Predicting the fatigue failure of glass/epoxy using cascade neural network with 20 neurons (Al-Assadi and El Kadi, 2009, with permission).

i. Obtain as many accurate experimental data as you can.

ANNs will give better results if trained using a large number of data points. ANNs will also respond positively if the scatter in the data is not significant. It should be emphasized that ANNs can accurately predict results within the range of the data used in training, but do not function very well when an extrapolating process is expected.

ii. Normalize the data.

Fatigue data can include values for the applied stress, fiber orientation angle, stress ratio, and number of cycles to failure among others. The ranges of values of these parameters vary; for example, the range for the fiber orientation angle is from 0° to 90°, whereas the range for the number of cycles to failure may be between 1 and 10^7 cycles. Training the network to learn such a wide range of values will produce unacceptable and unbalanced modeling performance since the ANNs will strive to minimize the overall MSE for all input patterns. Hence, minimizing the difference between the network output and observed data for the high number of cycles would be at the expense of the lower values.

iii. Choose your input parameters.

The predictions obtained using the ANNs will significantly depend on the input parameters fed to the network. One should experiment with the various combinations of parameters to find the set of input parameters leading to the optimum predictions.

iv. Choose an appropriate ANN architecture.

The most common ANN architecture is the FNN. It is suggested that any initial prediction should use this type of network. In cases where the FNN does not lead to accurate predictions, the MNN has been shown to lead to accurate predictions of fatigue life in composites. Other ANN structures have had less success in this area.

v. Choose the number of hidden layers and the number of neurons per hidden layer.

Always start with the simplest possible network, that is, one that contains one hidden layer. Considering the use of additional hidden layers should only be attempted if the relationship between the input and output parameters is very complicated. Using more than two hidden layers is not recommended. Although there is no exact method to select the number of neurons per hidden layer, for a neural network with one hidden layer, the number of neurons in the hidden layer should typically be between 3 and 4 times the number of input parameters. For a neural network with two hidden layers, the number of neurons in the first hidden layer should be between 3 and 4 times the number of inputs, whereas the number of neurons in the second hidden layer should be equal to the number of inputs to the ANN.

8.5   Polynomial Classifiers

In the previous section, ANNs were shown to give good fatigue life predictions in composite materials. To obtain such accurate predictions, many a priori decisions and parameters are needed such as the type of network better suited for a particular application, the type of algorithm used for training, the number of hidden layers to be used, or the number of neurons per hidden layer. Additionally, since the ANN formulation is of iterative type, the initial guess will no doubt influence the final result obtained. One should perform numerous runs and report the average results to be confident of the repeatability of the predictions obtained. For the above-mentioned reasons, a method requiring less a priori decisions while leading to repeatable results with less computational requirements should be sought.

The polynomial classifiers (PCs) are learning algorithms proposed and adopted in recent years for classification, regression, and recognition with remarkable properties and generalization ability (Campbell et al., 2002; Fukunaga, 1990; Devroye et al., 1996). Due to their need for less training examples and far less computational requirements, PCs are used in this work for composite life predictions. In the training phase, the elements of each training feature vector, $x = [\theta, \log(\sigma_{max})..., R]$, are combined with multipliers to form a set of basis functions, $p(x)$. The elements of $p(x)$ are the monomials of the form:

$$\prod_{j=1}^{N} x_j^{k_j}, \text{where } k_j \text{ is a positive integer and } o \le \sum_{j=1}^{N} k_j \le K \qquad (8.13)$$

For example, if vector x consists of two coefficients, $x = [\theta, \log(\sigma_{max})]$, and a second-degree polynomial (i.e., $K = 2$) is chosen, then:

$$p(x) = \begin{bmatrix} 1 & \theta & \log \sigma_{max} & \theta^2 & \theta \cdot \log \sigma_{max} & (\log \sigma_{max})^2 \end{bmatrix}^T \qquad (8.14)$$

Once the training feature vectors are expanded into their polynomial basis terms, the polynomial network is trained to approximate an ideal output using MSE as the objective criterion. The polynomial expansion for all of the training set features' vectors (*L* vectors) is defined as:

$$M = \begin{bmatrix} p(\theta) & p(\log \sigma_{max}) & \cdots & p(R) \end{bmatrix}^T \qquad (8.15)$$

The training problem reduces to finding an optimum set of weights, w, that minimizes the distance between the ideal outputs and a linear combination of the polynomial expansion of the training data such that (Al-Assadi and El Kadi, 2009):

$$w^{opt} = \arg_w \min \|Mw - O\|_2 \qquad (8.16)$$

where O represents the ideal output comprising of the column vector whose entries are the number of cycles to failure of the composite material under consideration.

The weights of the identification models, w^{opt}, can be obtained explicitly by applying the normal equations method (Al-Assadi and El Kadi, 2009) such as

$$w^{opt} = M^+ O \qquad (8.17)$$

where $\mathbf{M}^+$ is the Moore-Penrose pseudoinverse of matrix $\mathbf{M}$ (Lee et al., 1999).

In the prediction stage, when an unknown feature vector, x, is presented to the network, the vector is expanded into its polynomial terms $p(x)$ and its associated logarithmic number of cycles to failure prediction is determined, such that

$$\log(N_f) = w^{opt} p(x) \tag{8.18}$$

The PCs were introduced as a possible tool to predict fatigue life in composite materials (Al-Assaf and El Kadi, 2007). Experimental data obtained for various fiber orientation angles and stress ratios were used. The MATLAB (MATLAB) environment and its associated toolboxes are utilized for constructing, training, and testing of the classifiers. The use of first- and second-order classifiers was investigated but the best results were obtained when using a first-order classifier supplemented by the addition of a higher-order term:

$$p(x) = \begin{bmatrix} 1 & \theta & R & \log \sigma_{max} & \theta \cdot (\log \sigma_{max})^3 \end{bmatrix}^T \tag{8.19}$$

When compared to predictions obtained using FNNs, the PCs were shown to provide adequate modeling between the input parameters (maximum stress, R ratio, fiber orientation angle) and the number of cycles to failure (Figure 8.15).

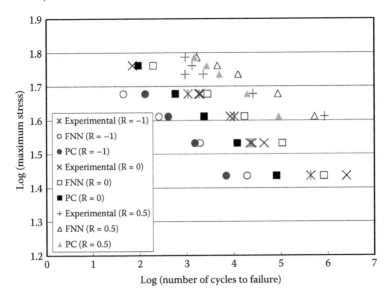

FIGURE 8.15
Comparison between fatigue life predictions using FNN and PC for $\theta = 45°$.

8.6 Conclusions

This work highlights the use of ANNs in the field of fatigue of composite materials. The sample detailed in this work shows that the use of ANNs can lead to predictions as accurate as, if not better, than those obtained by conventional methods. Developing an exact method of determining the appropriate ANN architecture, the number of hidden layers to use, and the number of neurons in each hidden layer are still topics for further research. A similar problem exists with PCs, where there is no exact method to determine which higher-order term will result in the best fatigue life prediction.

Group method of data handling is a multilayered network that might offer a solution to the above-mentioned drawbacks because it has the ability to determine which inputs are most relevant in determining fatigue life as well as to automatically determine the required number of hidden layers and the number of neurons in each layer. Each neuron within the network has two inputs and the function of the neuron is a second-order PC of its inputs. A regularity criterion determines which neuron should be rejected—in other words, which input combinations do not significantly affect the output. In addition, a closed form solution between the input and output parameters can be obtained. More information about group method of data handling can be found in the literature (Korbicz and Mrugalski, 2008; Kondo, 1998).

A significant advantage of neural networks would be their ability to predict the fatigue behavior of a laminate once the cyclic behavior of individual laminae is known. In addition to the variation of properties with the fiber orientation angle, this problem needs to incorporate the stacking sequence especially under fatigue loading where the sequence significantly alters the mechanical behavior of the laminae. The network would need to differentiate between the cyclic behavior of two laminates made from similar laminae stacked in a different order (e.g., $[90/0]_s$ vs. $[0/90]_s$).

References

Adam, T., Fernando, G., Dickson, R. F., Reiter, H., and Harris, B. 1989. Fatigue life prediction for hybrid composites. *International Journal of Fatigue* 11:233–237.

Akshantala, N. V., and Talreja, R. 2000. A micromechanics based model for predicting fatigue life of composite laminates. *Materials Science and Engineering A* 285:303–313.

Al-Assadi, M., and El Kadi, H. 2009. Predicting the fatigue life of different composite materials using artificial neural networks. *7th International Conference on Composite Science and Technology*, UAE.

Al-Assaf, Y., and El Kadi, H. 2001. Fatigue life prediction of unidirectional glass fiber/epoxy composite laminae using neural networks. *Composite Structures* 53:65–71.

Al-Assaf, Y., and El Kadi, H. 2007. Fatigue life prediction of composite materials using polynomial classifiers and recurrent neural networks. *Composite Structures* 77:561–569.

Aleksander, I., and Morton, H. 1990. *An introduction to neural computing.* London: Chapman & Hall.

Awerbuch, J., and Hahn, H. T. 1981. Off-axis fatigue of graphite/epoxy composites. In *Fatigue of fibrous composite materials,* ed. K. N. Lauraitis, pp. 243–273. ASTM STP 723.

Aymerich, F., and Serra, M. 1998. Prediction of fatigue strength of composite laminates by means of neural network. *Key Engineering Materials* 144:231–240.

Bishop, C. M. 1995. *Neural networks for pattern recognition.* Oxford: Oxford University Press.

Campbell, W., Assaleh, K., and Broun, C. 2002. Speaker recognition with polynomial classifiers. *IEEE Transactions on Speech and Audio Processing* 10:205–212.

Choi, S. W., Song, E.-J., and Hahn, H. T. 2003. Prediction of fatigue damage growth in notched composite laminates using an artificial neural network. *Composites Science and Technology* 63:661–675.

Devroye, L., Gyorfi, L., and Lugosi, G. 1996. *A probabilistic theory of pattern recognition.* New York: Springer-Verlag.

El Kadi, H. 2006. Modeling the mechanical behavior of fiber-reinforced polymeric composite materials using artificial neural networks—a review. *Composite Structures* 73:1–23.

El Kadi, H., and Al-Assaf, Y. 2002a. Energy-based fatigue life prediction of fiberglass/epoxy composites using modular neural networks. *Composite Structures* 57:85–89.

El Kadi, H., and Al-Assaf, Y. 2002b. Prediction of the fatigue life of unidirectional glass fiber/epoxy composite laminae using different neural network paradigms. *Composite Structures* 55:239–246.

El Kadi, H., and Al-Assaf, Y. 2007. The use of neural networks in the prediction of the fatigue life of different composite materials. *16th International Conference on Composite Materials, Japan.*

El Kadi, H., and Ellyin, F. 1994. Effect of stress ratio on the fatigue of unidirectional glass fibre/epoxy composite laminae. *Composites* 25:917–924.

Ellyin, F., and El Kadi, H. 1990. A fatigue failure criterion for fiber reinforced composite laminae. *Composite Structures* 15:61–74.

Elman, J. L., 1990. Finding structure in time. *Cognitive Science* 14:179–211.

Epaarachchi, J. A., and Clausen, P. D. 2003. An empirical model for fatigue behavior prediction of glass fiber-reinforced plastic composites for various stress ratios and test frequencies. *Composites Part A* 34:313–326.

Fausett, L. V. 1994. *Fundamentals of neural networks.* New Jersey: Prentice Hall.

Fawaz, Z., and Ellyin, F. 1995. A new methodology for the prediction of fatigue failure in multidirectional fiber-reinforced laminates. *Composites Science and Technology* 53:47–55.

Freire, R. C. S., Jr., Neto, A. D. D., and de Aquino, E. M. F. 2005. Building of constant life diagrams of fatigue using artificial neural networks. *International Journal of Fatigue* 27:746–751.

Freire, R. C. S., Jr., Neto, A. D. D., and de Aquino, E. M. F. 2007. Use of modular networks in the building of constant life diagrams. *International Journal of Fatigue* 29:389–396.

Freire, R. C. S., Jr., Neto, A. D. D., and de Aquino, E. M. F. 2009. Comparative study between ANN models and conventional equations in the analysis of fatigue failure of GFRP. *International Journal of Fatigue* 31:831–839.

Fukunaga, K. 1990. *Introduction to statistical pattern recognition.* New York: Academic.

Hajela, P. 2002. Soft computing in multidisciplinary aerospace design—new directions for research. *Progress in Aerospace Sciences* 38:1–21.

Hashin, Z., and Rotem, A. 1973. A fatigue failure criterion for fiber reinforced materials. *Journal of Composite Materials* 7:448–464.

Haykin, S. S. 1998. *Neural networks: a comprehensive foundation,* 2nd ed. Upper Saddle River: Prentice Hall.

Huang, Z. M. 2001. Micromechanical life prediction for composite laminates. *Mechanics of Materials* 33:185–199.

Jang, J. S. R., Sun, C. T., and Mizutani, E. 1997. *Neuro-fuzzy and soft computing: a computational approach to learning and machine intelligence.* New Jersey: Prentice Hall.

Kartalopoulos, S. V. 1996. *Understanding neural networks and fuzzy logic: basic concepts and applications.* New York: IEEE Press.

Kawai, M. 2004. A phenomenological model for off-axis fatigue behavior of unidirectional polymer matric composites under different stress ratios. *Composites Part A* 35:955–963.

Kawai, M., and Suda, H. 2004. Effects of non-negative mean stress on the off-axis fatigue behavior of unidirectional carbon/epoxy composites at room temperature. *Journal of Composite Materials* 38:833–854.

Kohonen, T. 1984. *Self-organization and associative memory.* Berlin: Springer-Verlag.

Kondo, T. 1998. GMDH neural network algorithm using the heuristic self-organization method and its application to the pattern identification problem. *Proceedings of the 37th SICE Annual Conference, International Session Paper,* pp. 1143–1148.

Korbicz, J., and Mrugalski, M. 2008. Confidence estimation of GMDH neural networks and its application in fault detection systems. *International Journal of Systems Science* 39:783–800.

Lee, J. A., Almond, D. P., and Harris, B. 1999. The use of neural networks for the prediction of fatigue lives of composite materials. *Composites Part A: Applied Science and Manufacturing* 30:1159–1169.

Mathur, S., Gope, P. C., and Sharma, J. K. 2007. Prediction of fatigue lives of composites material by artificial neural network. *Proceedings of the SEM Annual Conference and Exposition on Experimental and Applied Mechanics,* 2:1017–1024.

MATLAB Neural networks toolbox of MathWorks. www.mathworks.com.

McCulloch, W. W., and Pitts, W. 1943. A logic calculus of the ideas imminent in nervous activity. *Bulleting of Mathematical Biophysics* 5:115–133.

NEUDESK. Neural computer sciences. www.neusciences.com.

NeuroSolutions. NeuroSolutions software, www.nd.com.

Petermann, J., and Plumtree, A. 2001. A unified fatigue failure criterion for unidirectional laminates. *Composites Part A* 32:107–118.

Pidaparti, R. M., and Palakal, M. J. 1993. Material model for composites using neural networks. Technical notes. *AIAA Journal* 31:1533–1535.

Plumtree, A., and Cheng, G. X. 1999. A fatigue damage parameter for off-axis unidirectional fiber-reinforced composites. *International Journal of Fatigue* 21:849–856.

Powell, M. J. D. 1985. Radial basis function for multi-variable interpolation—a review. In *Proceedings of IMA Conference on Algorithm for the Approximation of Function, and Data*. RMCS.

Rosenblatt, F. 1958. The perceptron: a probabilistic model for information storage and organization in the brain. *Psychological Reviews* 65:386–408.

Rosenblatt, F. 1962. *Principles of neurodynamics*. Washington: Spartan Books.

Schalkoff, R. J. 1997. *Artificial neural networks*. London: McGraw-Hill.

Sihn, S., and Tsai, S.W. 2005. Prediction of fatigue S–N curves of composite laminates by Super Mic-Mac. *Composites Part A: Applied Science and Manufacturing* 36:1381–1388.

Skapura, D. 1996. *Building neural networks*. New York: ACM Press.

Swingler, K. 1996. *Applying neural networks: a practical guide*. London: Academic Press.

Tennyson, R. C. 1986. Strength and fatigue life design procedures for composite structures. *Proceedings of the 8th Symposium of Engineering Applications of Mechanics*. National Research Council of Canada (NRCC), pp. 32–39.

Tennyson, R. C., Wharram, G. E., and Mabson, G. E. 1986. Final report on investigation of static failure and compressive fatigue of CF-18 graphite/epoxy laminates—Part 1: Experimental investigation. Defence Research Establishment Pacific, Research and Development Branch, Department of National Defense, Victoria, British Columbia.

Varvani-Farahani, A., Haftchenari, H., and Panbechi, M. 2007. An energy-based fatigue damage parameter for off-axis unidirectional FRP composites. *Composite Structures* 79:381–389.

Vassilopoulos, A. P., and Bedi, R. 2008. Adaptive neuro-fuzzy inference system in modelling fatigue life of multidirectional composite laminates. *Computational Materials Science*, 43:1086–1093.

Zhang, Z., and Friedrich, K. 2003. Artificial neural networks applied to polymer composites: a review. *Composites Science and Technology* 63:2029–2044.

9

Optimizing Neural Network Prediction of Composite Fatigue Life Under Variable Amplitude Loading Using Bayesian Regularization

M. I. P. Hidayat and P. S. M. M. Yusoff

CONTENTS

ABSTRACT Neural networks (NN) found its application in fatigue field, especially in fatigue life assessment of composite materials in recent years. The use of NN in the field of application also implies the necessity for optimizing the NN prediction of the composite fatigue life with respect to the presence of limited fatigue data available and fatigue condition of varying stress amplitudes. In the present chapter, optimizing NN prediction of fatigue life under variable amplitude loading (various stress ratio conditions) in relation with the availability of limited fatigue data for polymeric-based composites is presented. Multilayer perceptrons-based NN model is developed, and the training algorithm of Levenberg-Marquardt incorporating adaptive Bayesian regularization is used in the present study. From the simulation results obtained, it can be shown that training the developed network with fatigue data of only two stress ratios, which represent limited fatigue data, gave reasonably accurate fatigue life prediction under wide range of stress ratio values. The reliability and accuracy of the NN prediction were quantified by small mean square error (MSE) values. Finally, when using much less training fatigue data (22% from the total fatigue data), the network can still produce significant coefficient of determination between the prediction results and those obtained by the experiment.

9.1   Introduction

In design of composite material structures, fatigue failure is the most important aspect because it is closely related to performance, reliability, and durability of the structures (Reifsnider, 1991). It is a routine task done by design engineers to estimate the structure lifetime under long-term fatigue loading influence to ensure good reliability during its service. As a consequence, lifetime assessment of a component material becomes an important part in design phase to assess the useful life of the material-related components or structures.

The necessity of lifetime assessment in composite materials is also driven by the need to speed up the time frame from the research stage to the marketplace and is also influenced by the intensive use of composite materials that require performance assessment in the competitive market.

Particularly from a design engineer's point of view, fatigue life assessment for multivariable amplitude loading is the most challenging item. A real structure or component is exposed to an extremely complicated pattern of stress cycles of varying amplitude and mean stress along its service life. It means that fatigue life assessment of multivariable amplitude loading, a wide spectrum of varying amplitude, and mean stresses of loading need to be considered.

If covering such a wide varying-amplitude spectrum and mean stresses of loading in the fatigue life assessment is most concerned, performing fatigue testing at several different stress ratios (R) should be done. However, it is well-known that fatigue testing is very time-consuming and costly. Thus, fatigue life assessment of a material is frequently faced with tight compromises among the available time, effort, and cost. On the other hand, fatigue strength data at several stress ratios are not always available.

In the last decade, researchers in the fatigue field have explored neural networks (NN) as an information-processing and computing tool to deal with fatigue-related problems (Aymerich and Serra, 1998; Lee and Almond, 2003). The characteristic of NN that can be taught to emulate relationships in sets of data to subsequently predict the outcome of another new set of input data, for example, another composite system or a different stress environment, is exploited to yield faster acquisition of fatigue data, thus reducing experimentation time and cutting down the associated high costs. Moreover, NN becomes an alternative approach in fatigue life assessment because of its ability to handle data conveniently and ease of implementation as well as the wide availability of software environments to implement it.

9.1.1 Review on Fatigue Life Assessment of Composite Materials Using NN

NN found its application in fatigue field, especially in fatigue life assessment of composite materials under variable amplitude loading. The use of NN has been a new route in the fatigue life assessment, where the aim is to develop NN model(s) producing reliable prediction using a limited body of fatigue data that in turn support related design decisions immediately and reliably. The use of NN in fatigue life assessment of composite materials has a wide range of applications from unidirectional (Al-Assaf and El-Kadi, 2001; El-Kadi and Al-Assaf, 2002) to multidirectional laminate (Freire et al., 2005; Vassilopoulos, 2007; Freire et al., 2007). A comprehensive review of the recent works is presented here.

Al-Assaf and El Kadi (2001) and El Kadi and Al-Assaf (2002) assessed the fatigue life of unidirectional glass fiber/epoxy laminate using different NN paradigms, namely, feed-forward (FF), modular, radial basis function, and principal component analysis networks, and compared the prediction results with the experimental data. Standard back propagation algorithm was used in the study. Specimens with five fiber angle orientations of $0°$, $19°$, $45°$, $71°$, and $90°$ were tested under three stress ratio (R) conditions of -1, 0, and 0.5. Ninety-two experiment data made up the application data for the networks. They found that NN can be trained to model the nonlinear behavior of composite laminate subjected to cyclic loading, and the prediction results were comparable with other current fatigue-life prediction methods. Nevertheless, only three stress ratios (R) were investigated, and some large discrepancies in experimental data were still found.

Freire et al. (2005) followed a different approach, wherein NN was used to build constant life diagrams (CLDs) of fatigue. The researchers built CLDs with plastic reinforced with fiberglass (DD16 material) with $[90/0/\pm45/0]_s$ layup. Four training data sets (each set consists of 3R, 4R, 5R, and 6R values, respectively) were set up from 12 stress ratio (R) values. The training algorithm was the standard back propagation. It was found that the use of NN to build CLD was very promising where the NN trained using only three *S–N* curves could generalize and construct other remaining S–N curves of the CLD building. However, it was also pointed out that six S–N curves should be used in NN training for better generalization.

Vassilopoulos et al. (2007) criticized that the determination of six S–N curves was a costly task for the NN prediction purpose. Instead, these authors used a small portion, namely 40%–50%, of the experimental data. It was shown that it is possible to build CLD using the small portion data and NN was proven to be a sufficient tool for modeling fatigue life of GFRP multidirectional laminates. The authors, however, stated their NN model weakness to extrapolate data outside the training region as one of the major unresolved problems.

In their next article, Freire et al. (2007) showed that the use of modular networks gives more satisfactory results than FF NN. However, it was still necessary to increase the training sets for better results. Finally, Zhang and Friedrich (2003) stated that fatigue behavior is still so complicated that the problem requires more effort before NN can be used with more confidence.

From recent investigations, NN has been proven to be a sufficient tool for modeling fatigue life of composite materials, ranging from unidirectional to multidirectional laminate. In addition, NN also shows the potential for fatigue life assessment under variable amplitude loading. However, no attempts have been made so far to use less training data set and at the same time ensure reasonably accurate fatigue life prediction. In other words, it is necessary to optimize NN prediction of composite fatigue life under variable amplitude loading using less fatigue data to take full advantage of the NN potential for much more efficient fatigue life assessment.

9.1.2 Motivation and Objective

Previous studies still offer less satisfactory fatigue life prediction results in the sense that more training sets are needed for better prediction, in particular, for fatigue life assessment under variable amplitude loading. The current study is in the spirit of how a limited body of fatigue data could be optimized using an NN model and provides faster (but still accurate) support for design decisions.

Bearing the motivation in mind, the main objective of the current study is to build an efficient multilayer perceptrons (MLP)-based NN model and use it to predict fatigue lives of polymeric-based composites under variable amplitude loading or different stress ratios condition using and optimizing limited fatigue data as the training data set.

To achieve the objective, the authors have defined a set of stress ratios that consists of only two R values as a limited stress ratio set, and the utilization of fatigue data from the limited set was investigated. This is in contrast with the previous studies that used fatigue data from at least three R values as training set to obtain reasonable fatigue life predictions. In addition, a form of regularization, namely Bayesian regularization, was used to deal with the utilization of the limited fatigue data as training data set in the NN modeling. It is well-known that NN modeling with only a few training examples leads to an ill-posed problem, and regularization is needed to avoid frequent trapping of NN weights in local minima (Bishop, 1995). Moreover, regularization is implemented within the training algorithm of Levenberg-Marquardt (LM) in an adaptive manner without the need to fix the regularization parameters manually. The training algorithm was chosen because it has rapid convergence and robustness properties in solving optimization problem (Fletcher, 1980).

The next sections are organized as follows: Section 9.2 presents the important aspects of fatigue life assessment under variable amplitude loading related to the current study. In Section 9.3, problem formulation of NN learning is presented, and some aspects related to NN optimization and generalization are described briefly. Materials and methods are presented in Section 9.4, and Section 9.5 follows with the NN model fatigue life prediction results together with the related comparison and discussion of experimental data. For better insights, the comparisons are also presented in the form of the classical S–N curve. Section 9.6 closes with conclusions and recommendations for further studies.

9.2 Fatigue Life Assessment Under Variable Amplitude Loading

The fact that a component or structure is frequently subjected to complex stress cycle patterns of varying amplitudes and mean stresses along its service life needs a suitable representation that is able to describe the condition efficiently and effectively.

CLD is a very useful and convenient tool in fatigue life assessment under variable amplitude loading. In a CLD, the points along each radial line are the points of S–N curve for a specific stress ratio (R), and all points with the same fatigue life are connected with lines in a plane of amplitude stress (σ_a) – mean stress (σ_m) axes. It is clear that CLD is another way to represent the S–N curve, which is very familiar to design engineers. The schematic of CLD is shown in Figure 9.1.

The relationship between stress amplitude (σ_a) and mean stress (σ_m) for each of the radial lines is described in equation (9.1):

$$\sigma_a = \left(\frac{1-R}{1+R}\right)\sigma_m \tag{9.1}$$

except for $R = -1$ where $\sigma_m = 0$ and $\sigma_a = \sigma_{max}$.

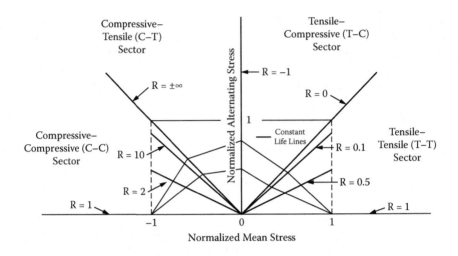

FIGURE 9.1
A schematic representation of CLD.

Looking at the CLD, three different fatigue loading regions or sectors of tension–tension (T–T), tension–compression (T–C), and compression–compression (C–C) of fatigue are represented clearly. Different conditions of fatigue with variable mean and amplitude stresses having the same lifetime as well as the distribution of the conditions then could be identified easily using the diagram.

From the description of CLD, the authors want to point out how the definition of a limited set of stress ratios found its concept from the CLD, as described in what follows. A limited set of stress ratios could be categorized as well-distributed or not well-distributed in CLD regions. Recall that in Figure 9.1, a set of stress ratios is said to be well-distributed in CLD regions if the stress ratios are symmetrical in position to one another. Otherwise, it is said to be not well-distributed. For instance, a set of stress ratios of 0.1 and 10 or 0.5 and 2 is said to be well-distributed in the CLD because the pair of the stress ratios are symmetrical in position to one another. On the other hand, a set of stress ratios of 0.1 and 0.5 is said to be not well-distributed.

The definition of limited stress ratios set will be used throughout the current study, and the NN model capability to predict fatigue lives at other stress ratio values using the limited set of stress ratios, whether well-distributed or not, as training set is to be further examined.

9.3 NN Design

The NN architecture of MLP with one hidden layer was used to perform the fatigue life modeling task in the current study. The architecture was chosen

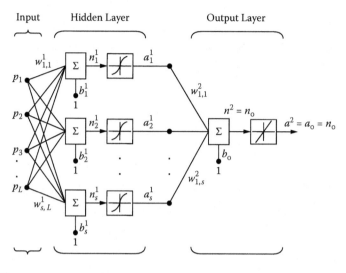

FIGURE 9.2
MLP with one hidden layer and single output.

because its design is relatively simple and straightforward. Also, various training algorithms have been established and are available to be chosen for such NN architecture (Hagan and Demuth, 1996). Figure 9.2 shows an MLP with one hidden layer and single output.

The notations presented in Figure 9.2 are p (input sets), L (number of elements in input vector), s (number of hidden nodes), n (summed up of weighted inputs), a (output of activation function in the corresponding layer, $w^1_{j,i}$ and b^1_j (input weight and bias; $i = 1$ to L, $j = 1$ to s), $w^2_{1,j}$ and b_o (layer weight and output bias), and y (MLP output). Superscripts 1 and 2 represent the first layer of hidden and the second layer of output, respectively.

Learning in NN is achieved by adjusting the corresponding weights in response to external environment of input sets. The weights adjustment is accomplished by a set of learning rule by which an objective function is minimized. In what follows, problem formulation of NN learning will be concisely presented, particularly from the supervised learning context of the MLP.

Let (P,T) be a pair of random variables with values in $P = \Re^m$ and $T = \Re$, respectively. The regression of T on P is a function of P, f: $P \circledR T$, giving the mean value of T conditioned on P, $E(T|P)$.

Let random samples $O_1^Q = \left\{ (P_1, T_1), \ldots, (P_Q, T_Q) \right\}$ of size Q be drawn from the distribution of (P,T) as an observation set. For $Q \geq 1$, $\hat{f}_Q$ will denote an estimator of f based on the random samples, that is, a map $\hat{f}_Q$: $O_1^Q \to \hat{f}_Q\left(O_1^Q, . \right)$, where, for fixed O_1^Q, $p \to \hat{f}_Q\left(O_1^Q, p \right)$ is an estimate of the regression function $f(p)$.

Furthermore, for convenience, P and T will be referred to as the sets of input and variable output, respectively. Given the observation set O, learning in NN for realization of the estimate $\hat{f}$ means adjusting to vector of parameters weight **w** and biases **b** using a set of learning rule or learning algorithm in such a way that $\hat{f}$ minimizes the objective function or empirical error defined as

$$E(\mathbf{w}) = \sum_{q=1}^{Q} \left[t_q - \hat{f}\left(\mathbf{p}_q; \mathbf{w}\right) \right]^2 \tag{9.2}$$

and generalizes well or outputs properly when a novel input vector $\mathbf{p}_{\text{test}}$ never seen before is fed into the network.

The estimate $\hat{f}$ realized by the MLP shown in Figure 9.2 given the training set O can be written as

$$\hat{f}(\mathbf{p}; \mathbf{w}) = \sum_{i=1}^{s} w_{1,i}^2 \, \tau\left(w^1_{i,j} p + b_i\right) + b_0 \tag{9.3}$$

where $\tau(.)$ is a sigmoidal function used in the nodes of hidden layer.

In addition, by keeping in mind that learning in NN is principally updating the network weights based on the given set of examples so that the network will give proper response to new examples, below are two limiting factors of the NN learning.

First, only a finite number of observation points (example pairs) are available. This means that the available examples sometimes must be fully used for the NN learning purpose to provide proper learning of the underlying process. Hence, the practicability and feasibility of using limited examples for NN learning to yield accurate prediction output are assured.

The second is that the realization of target at the points of observation p_q, $q = 1, \ldots, Q$ is observed with an additive noise e_q:

$$e_q = T_q - f(P_q) \tag{9.4}$$

The observations are then noisy, and the target noises e_q introduce a random component in the estimation error.

9.3.1   Optimization and Gradient-Based Learning Methods

The NN learning process by which the NN weights are updated is, in principle, a minimization process to particular objective function, and gradient-based learning methods are commonly used for the NN learning (Haykin, 1994). In this section, gradient-based learning methods are described and introduced from the minimization statement, leading to the derivation of the LM formula for updating the NN weights.

Suppose that the function to be minimized is of the following special form given by Equation (9.5):

$$f(\mathbf{x}) = \frac{1}{2} \sum_{j=1}^{m} r_j(\mathbf{x})^2 \tag{9.5}$$

where $\mathbf{x} = x_i = (x_1, x_2, \ldots, x_n)$ is a vector where $1 \leq i \leq n$ and each of r_i is a function from $\mathfrak{R}^n$ to $\mathfrak{R}$, where $1 \leq j \leq m$. r_j is referred to as a *residual*.

Residual vector $\mathbf{r}$ is a vector function of $\mathbf{x}$: $\mathfrak{R}^n$ to $\mathfrak{R}^m$, hence $f(\mathbf{x})$ can be rewritten as

$$f(\mathbf{x}) = \frac{1}{2} \|\mathbf{r}\|^2 \tag{9.6}$$

The Jacobian matrix $\mathbf{J}$ of r with respect to $\mathbf{x}$ can be defined as

$$\mathbf{J} = \frac{\partial r_j(\mathbf{x})}{\partial \mathbf{x}} \tag{9.7}$$

Gradient $\mathbf{g}$ and Hessian $\mathbf{H}$ are, respectively, defined in Equations (9.8) and (9.9) as

$$\mathbf{g} = \nabla f(\mathbf{x}) = \sum_{j=1}^{m} r_j(\mathbf{x}) \frac{\partial r_j(\mathbf{x})}{\partial \mathbf{x}} = \mathbf{J}^{\mathrm{T}} \mathbf{r} \tag{9.8}$$

$$\mathbf{H} = \nabla^2 f(\mathbf{x}) = \mathbf{J}^{\mathrm{T}} \mathbf{J} + \sum_{j=1}^{m} r_j(\mathbf{x}) \nabla^2 r_j(\mathbf{x}) \tag{9.9}$$

If it is possible to approximate the *residuals* r_j by linear functions ($\nabla^2 r_j(\mathbf{x})$ are small) or the *residuals* r_j themselves are small, the Hessian $\mathbf{H}$ can be approximated as

$$\mathbf{H} = \nabla^2 f(\mathbf{x}) = \mathbf{J}^{\mathrm{T}} \mathbf{J} \tag{9.10}$$

Further, if vector $\mathbf{x}$ represents weights vector $\mathbf{w}$ and $f(\mathbf{x})$ represents the performance function $E(\mathbf{w})$, the basic weights update rule, which is a gradient descent rule, in NN training can be written as

$$\mathbf{w}_{i+1} = \mathbf{w}_i - \eta \nabla E(\mathbf{w})\big|_i \qquad (9.11)$$

$$\mathbf{w}_{i+1} = \mathbf{w}_i - \eta \mathbf{g}\big|_i \qquad (9.12)$$

where $\mathbf{w}_{i+1}$ is weights vector at next iteration, $\mathbf{w}_i$ is weights vector at current iteration, η is learning rate factor, and i is iteration step.

Thus, in NN training, the weights are moved along the negative of the performance function gradient $\mathbf{g}$ using the gradient descent learning.

Equation (9.12) can be considered as a first-order approximation to gradient descent, and therefore, the method is often very slow in practice. To overcome the weakness of basic gradient descent, methods based on second-order approximation to gradient descent have been developed, such as the Newton and LM methods.

If $E(\mathbf{w})$ is expanded using a Taylor series in minimizing the performance function, then the Newton's method for updating weights will be found as shown in Equation (9.13).

$$\mathbf{w}_{i+1} = \mathbf{w}_i - (\nabla^2 E(\mathbf{w}))^{-1} \nabla E(\mathbf{w})\big|_i \qquad (9.13)$$

Equation (9.13) can also be written in the form of equation (9.14):

$$\mathbf{w}_{i+1} = \mathbf{w}_i - (\mathbf{H})^{-1} \mathbf{g}\big|_i \qquad (9.14)$$

The LM method for updating weights is by adding an adjustable constant parameter λ to Equation (9.14) to become

$$\mathbf{w}_{i+1} = \mathbf{w}_i - (\mathbf{H} + \lambda \mathbf{I})^{-1} \mathbf{g}\big|_i \qquad (9.15)$$

where λ is the lambda parameter or the parameter of LM.

With the presence of the adjustable parameter λ, the LM algorithm combines both the advantages of the simple gradient descent that simply moves the weights to decrease the error function value and the Newton's method that has faster convergence. In practice, for moderate-size problems, the LM algorithm is faster and robust. The LM method also outperforms the methods because $E(\mathbf{w})$ is always reduced at each iteration of the algorithm. For further reading on optimization, the readers are directed to Fletcher (1980) and Nocedal and Wright (2006).

9.3.2 Bayesian Regularization

Discovering the function f, or an estimate of it $(\hat{f})$, from the observation data O given is basically an ill-posed problem, as the estimate can have infinite solutions, particularly if the nature of the function f is hard to be assessed.

To help in choosing one particular preferred solution from the infinitely many solutions, a priori information or knowledge of the function f is needed. For example, function f is assumed to be smooth, in the sense that two similar inputs will produce two similar outputs, or by assuming the additive noise of equation (9.4) is drawn from Gaussian distribution.

Regularization is then a way to prevent overfitting, a condition where good generalization is not likely to occur. Bayesian regularization was used in this study to improve the generalization capability, thus the quality of NN prediction. It minimizes a linear combination of squared errors and weights. The idea is to find a balance between the number of parameters and goodness of fit by penalizing large models (Bishop, 1995).

Incorporating the regularization, the cost function of equation (9.2) is then modified as follows:

$$E(\mathbf{w}) = \beta \sum_{q=1}^{Q} \left[t_q - \hat{f}\left(\mathbf{p}_q; \mathbf{w}\right) \right]^2 + \alpha \sum_{i=1}^{W} w_i^2 \qquad (9.16)$$

where α is a weight decay parameter, β is an inverse noise variance parameter, and W is the total number of weights. The noise is assumed to apply to the target data t_q and its distribution is assumed to be a zero-mean Gaussian distribution.

Note that the new cost function consists of the sum of squared errors (E_D) and the sum of squared weights (E_w) terms. In addition, parameters α and β are introduced for penalizing large models. Using the modified cost function, it is clear that there is a need to reestimate the parameters accordingly. One estimation method is the Gauss-Newton approximation implemented to the Bayesian learning within the framework of the LM algorithm (Foresee and Hagan, 1997).

Using the form of regularization, an NN model with fewer number of parameters (weights) is preferred than the one with large number of weights, especially for small data sets or training examples to yield a smoother and stable NN response. In the current study, the consistency of the NN prediction with respect to limited training examples of fatigue data will be shown later in the Simulation Results and Discussion section.

9.4　Materials and Methods

9.4.1　Materials

The investigated materials were E-glass fabrics/epoxy and E-glass/polyester, typical materials used in wind turbine blade applications. Named as QQ1 and DD16, respectively, with the corresponding layups of $[\pm45/0_4/\pm45/]$ and $[90/0/\pm45/0]_S$, the materials represent the current and the earlier materials used in the applications. The QQ1 has a higher fiber content than the DD16 and is more representative of the materials being used today. Moreover, the materials have been studied extensively in the past (DOE/MSU, 2007; Mandell and Samborsky, 1997). The corresponding database containing fatigue data of various R values ($R = 0.1, 0.5, -0.5, -1, -2$, and 10 for E-glass/epoxy and $R = 0.1, 0.5, 0.7, 0.8, 0.9, -0.5, -1, -2$, and 10 for E-glass/polyester) makes it suitable for the study purpose.

From the fatigue data, stress ratio (R), maximum stress (S_{max}), and minimum stress (S_{min}) values were used as input set, and the output was the corresponding fatigue cycles (log N) for the input set. For each particular R value, mean fatigue life values were used. All the data were normalized in the range of -1 to 1.

Furthermore, several sets of the stress ratios were examined and evaluated to seek the best training set from which better fatigue life prediction results could be achieved. In turn, knowing the best training set would be very useful in understanding how NN responds to this fatigue life prediction problem and which stress ratios might be appropriate to be selected and collected first.

In the current study, three combinations of stress ratios were determined and selected for the composites. For E-glass/epoxy, the combinations of stress ratios used as training sets were $R = 0.1$ and 0.5 (T–T and T–T), $R = 0.5$ and -0.5 (T–T and T–C), and $R = 0.1$ and -2 (T–T and T–C). Meanwhile, for E-glass/polyester, the training sets were $R = 0.1$ and 0.5 (T–T and T–T), $R = 0.1$ and -1 (T–T and T–C), and $R = 0.1$ and 10 (T–T and C–C). The combinations of stress ratios are summarized in Table 9.1.

The rationale of using the combinations of stress ratios is twofold. First, the selected combinations of stress ratios represented two halves sector of CLD as shown in Figure 9.1. Thus, the selections represent different representative

TABLE 9.1

Combinations of Stress Ratios Used as Training Sets

Material	Training Sets		
	Set I	Set II	Set III
E-glass/epoxy	$R = 0.1$ and 0.5	$R = 0.5$ and -0.5	$R = 0.1$ and -2
E-glass/polyester	$R = 0.1$ and 0.5	$R = 0.1$ and -1	$R = 0.1$ and 10

conditions of fatigue state and capture the behavior of fatigue data through-out the CLD regions to give better prediction results for the NN modeling. Second, failure mechanisms of fatigue under T–T condition are different from the ones under C–C condition. Failures of fiber breaking, interfacial debonding, and fiber pull-out play important roles in fatigue under T–T; meanwhile, fiber kinking, buckling, and fiber–matrix shear-out dominate in fatigue under C–C. Therefore, the selections were chosen to examine the NN modeling effectiveness in constructing the CLD regions. Figure 9.3 shows the steps involved in the NN modeling of the fatigue life assessment.

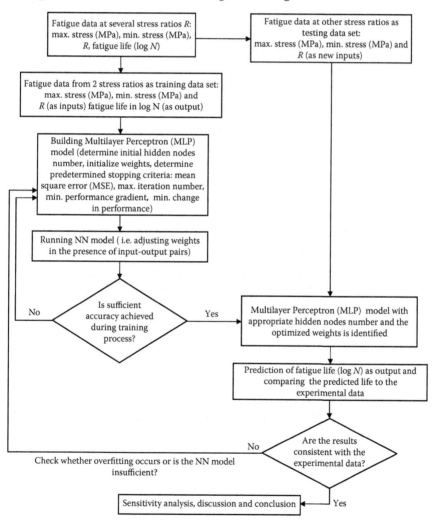

FIGURE 9.3
Flowchart of the NN modeling process in the current study.

9.4.2   Adaptation of Bayesian Framework Within the LM Algorithm

In principle, adaptation of Bayesian framework means that the inclusion of updating parameters α and β within the LM steps (Foresee and Hagan, 1997).

Recall equation (9.16), which is rewritten here as

$$E(\mathbf{w}) = \beta E_D + \alpha E_w \tag{9.17}$$

where

$$E_D = \sum_{q=1}^{Q}\left[t_q - \hat{f}\left(\mathbf{p}_q;\mathbf{w}\right)\right]^2 \tag{9.18}$$

$$E_w = \sum_{i=1}^{W} w_i^2 \tag{9.19}$$

Using the modified cost function, the gradient $\mathbf{g}$ and Hessian $\mathbf{H}$, respectively, become

$$\mathbf{g} = 2\beta\mathbf{J}^{\mathrm{T}}\mathbf{r} + 2\alpha\mathbf{w} \tag{9.20}$$

$$\mathbf{H} = 2\beta\mathbf{J}^{\mathrm{T}}\mathbf{J} + 2\alpha\mathbf{I} \tag{9.21}$$

Thus, the increment of weights $\Delta\mathbf{w}$ becomes

$$\Delta\mathbf{w} = -\left[\beta\mathbf{J}^{\mathrm{T}}\mathbf{J} + (\lambda+\alpha)\mathbf{I}\right]^{-1}\mathbf{g} \tag{9.22}$$

Furthermore, for the purpose of updating α and β, the Hessian formulation was used through the following equations:

$$\gamma = I - 2\alpha\,\mathrm{trace}\left(\mathbf{H}^{-1}\right) \tag{9.23}$$

$$\alpha = \frac{\gamma}{2E_w} \tag{9.24}$$

$$\beta = \frac{Q-\gamma}{2E_D} \tag{9.25}$$

where γ is the effective number of parameters and is a measure of how many parameters or weights are effectively used (preferred) in the NN learning with respect to the cost function reduction, I is the total number of initial weights during initialization, and Q is the number of training examples.

Based on the above formulas, the LM algorithm implementing Bayesian regularization can be stated as follows:

Step 1: The weights **w** and parameters λ, α, and β were initialized. For example, $\lambda = 0.005$, $\alpha = 0$, and $\beta = 1$. The algorithm is not too sensitive to the initial choice of the parameters. In addition, the choice of $\alpha = 0$ and $\beta = 1$ means that the NN is starting from the original cost function. Recall Equation (9.16).

Step 2: One step of the LM algorithm to minimize the objective function was taken as per Equation (9.22).

Step 3: If $E(\mathbf{w} + \Delta\mathbf{w}) < E(\mathbf{w})$, then $\mathbf{w}_{new} = \mathbf{w} + \Delta\mathbf{w}$ was accepted as a new iteration.

Step 4: The effective number of parameter γ was computed using equation (9.23), and the Hessian formulation of Equation (9.21) was used.

Step 5: The parameters α and β were updated using Equation (9.24) for α and Equation (9.25) for β.

Step 6: Steps 2–5 were repeated until the stopping criterion was satisfied or convergence was achieved.

9.4.3 Realization

Programming lines have been written in MATLAB to implement the above pseudocodes.

The MATLAB programming lines were run on Toshiba Satellite with OS Windows Vista Basic, processor of Intel Pentium Dual-Core, and RAM of 1 GB.

9.5 Simulation Results and Discussion

The consistency and reliability of NN fatigue life prediction using the LM algorithm for the polymeric composites will be presented in this section by comparing the predicted results with the corresponding experimental or simulated data. For better insights, the comparisons are also presented in the form of the classical *S–N* curve. Using the algorithm scheme, the associated NN regularization parameters α and β are adjusted automatically and

TABLE 9.2

NN Parameters Used in the Modeling Study

NN Parameter	Value
Initial lambda, λ_{init}	0.005
Initial weight decay, α_{init}	0
Initial inverse noise, β_{init}	1
Maximum number of iterations	200
Minimum gradient, g_{min}	1×10^{-10}
Maximum lambda, λ_{max}	1×10^{10}
Performance goal	0
Number of hidden nodes, s	10

updated iteratively during the NN iterations. Discussions on the results are then presented accordingly. Finally, the quality of the NN fatigue life prediction is assessed systematically.

9.5.1   NN Fatigue Life Prediction

The following fatigue life prediction results have been collected based on the NN parameters and stopping criteria listed in Table 9.2.

9.5.1.1   *Material I (E-Glass/Epoxy, [±45/0₄/±45/])*

The results of the NN fatigue life prediction for material I (E-glass/epoxy, [±45/0$_4$/±45/]) are presented. For this composite, fatigue data from three sets of stress ratios of 0.1 and 0.5, 0.5 and −0.5, and 0.1 and −2 were used as the training data sets. The selection of stress ratios represented the different conditions of fatigue state. The NN fatigue life prediction results are shown in Figures 9.4 to 9.6.

From Figures 9.4 to 9.6, it can be seen that the NN prediction of the material's fatigue lives are consistent with that of the experimental values. Also, the NN prediction trends closely match the experimental data trends. This means that the NN model is able to capture the necessary information and general trend of the fatigue data well during its learning phase.

The results also showed that there is no significant difference in the precision of the prediction as the training sets are varied from T–T to T–C condition mode. This is evidenced by consistent MSE values calculated for the three different training sets, as shown in Table 9.3. On average, the MSE measured for material I is 0.12 ± 0.01. Based on the findings, the use of the training set of T–T stress ratios may be preferred compared with the training set of T–C or C–C stress ratios due to the simplicity of T–T fatigue testing. Fatigue testing under high-compression condition may require the use of additional guide and supports to avoid buckling of the sample.

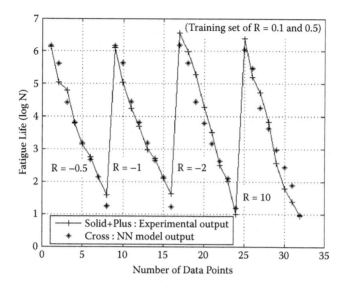

FIGURE 9.4
Fatigue life prediction by NN model for testing sets $R = -0.5, -1, -2,$ and 10 using $R = 0.1$ and 0.5 as training set.

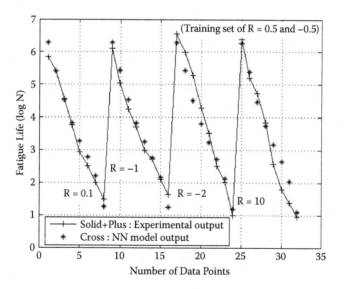

FIGURE 9.5
Fatigue life prediction by NN model for testing sets $R = 0.1, -1, -2,$ and 10 using $R = 0.5$ and -0.5 as training set.

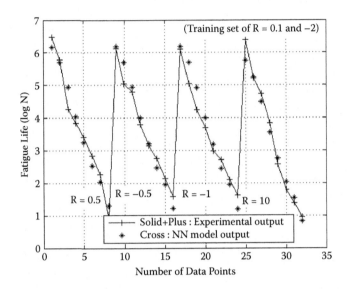

FIGURE 9.6
Fatigue life prediction by NN model for testing sets $R = 0.5$, -0.5, -1, and 10 using $R = 0.1$ and -2 as training set.

TABLE 9.3

The MSE of the Fatigue Life Prediction Results of Material I for Various Training Sets

Training Set	R Values	MSE Prediction Value
Set A	0.1 (T–T) and 0.5 (T–T)	0.12
Set B	0.5 (T–T) and –0.5 (T–C)	0.13
Set C	0.1 (T–T) and –2 (T–C)	0.11

From Table 9.3, it is observed that the training set of 0.1 and –2 gives the best prediction results indicated by the lowest MSE value of 0.11. Although both training sets B and C are of the combination T–T and T–C mode, training set C gave better prediction results with lower MSE value. This may be because of the wider spread of fatigue data as demonstrated in the CLD.

For better insights, the fatigue life prediction results are also displayed in the classical *S–N* curve form as shown in Figures 9.7 to 9.10. The results presented are for those using $R = 0.1$ and -2 as the training set with the lowest MSE. Similar observations apply to training sets A and B.

The *S–N* curves predicted by the NN model correlate well with those based on experimental data for all the stress ratios under study as indicated by the high coefficient of determination r^2 values ranging from 0.9526 to 0.9809. Note that the discrepancies between the experimental data and the NN pre-

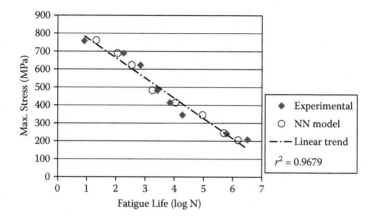

FIGURE 9.7
The *S–N* curves based on experimental data and as predicted by the NN model for *R* = 0.5.

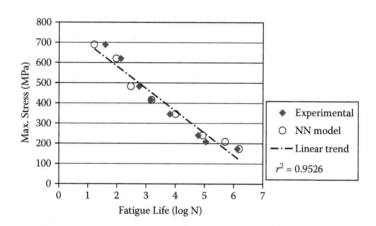

FIGURE 9.8
The *S–N* curves based on experimental data and as predicted by the NN model for *R* = −0.5.

dicted values fall within reasonable range. The general trend observed was also justifiable.

Note that there were eight fatigue data involved at each stress ratio. In particular, for material I, there was a total of 48 fatigue data used in the training and testing phases at six different stress ratios. As two stress ratios were used as the training set in predicting the fatigue life at other four *R* values, this represents 33% utilization of the data as training examples. The utilization of the limited fatigue data as the NN training examples illustrates the value of the current study.

One of the most important findings from the current study is the possibility to assess and predict the fatigue life of composites at the beginning of

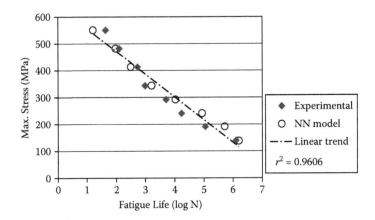

FIGURE 9.9
The *S–N* curves based on experimental data and as predicted by the NN model for $R = -1$.

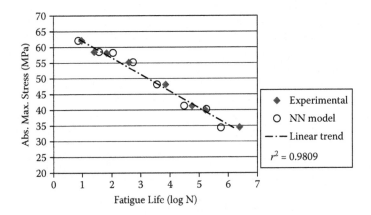

FIGURE 9.10
The *S–N* curves based on experimental data and as predicted by the NN model for $R = 10$.

the design or testing stage with limited fatigue data available. It has been shown that reliable and consistent prediction of fatigue life was achieved using fatigue data at only two *R* values as the training set.

9.5.1.2 Material II (E-Glass/Polyester, [90/0/±45/0]s)

In this section, the results of the NN fatigue life prediction for material II (E-glass/polyester, [90/0/±45/0]ₛ) are presented. Different from material I, where eight fatigue data were involved at each stress ratio, there were only five fatigue data at each stress ratio for this material. However, more *R* values are involved than those of material I. It can be said that the NN prediction of the

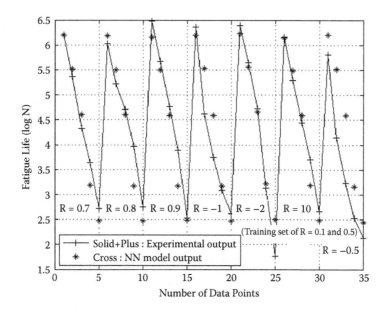

FIGURE 9.11
Fatigue life prediction by NN model for testing sets R = 0.7, 0.8, 0.9, −1, −2, 10, and −0.5 using R = 0.1 and 0.5 as training set.

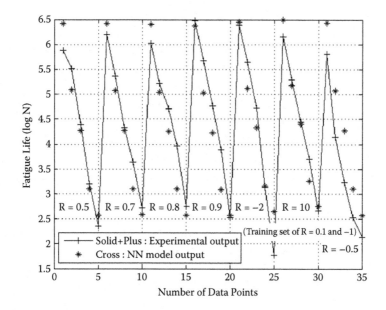

FIGURE 9.12
Fatigue life prediction by NN model for testing sets R = 0.5, 0.7, 0.8, 0.9, −2, 10, and −0.5 using R = 0.1 and −1 as training set.

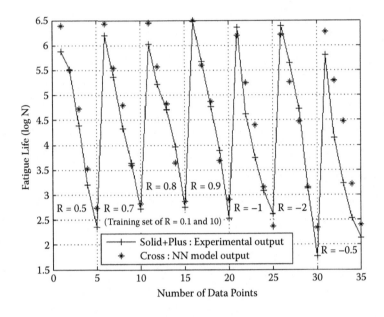

FIGURE 9.13
Fatigue life prediction by NN model for testing sets $R = 0.5$, 0.7, 0.8, 0.9, −1, −2, and −0.5 using $R = 0.1$ and 10 as training set.

material's fatigue life is more challenging with lesser training examples. The NN fatigue life prediction results obtained are shown in Figures 9.11 to 9.13.

As pointed out in Table 9.2, the NN model used to predict the fatigue life of all the three materials used the same number of hidden nodes of 10; thus, the prediction results are comparable from material to material. For material II, however, the number of fatigue data involved at each stress ratio is 5, which is lesser than the ones for material I as can be seen in Figures 9.11 to 9.13. With nine stress ratios involved in total and with only two used as the training set, the percentage of the training samples was 22% of the total fatigue data.

From Figures 9.11 to 9.13, it is interesting that with lesser training examples, the fatigue life prediction results are generally still consistent with experimental data, with an exception for fatigue life prediction at stress ratio $R = -0.5$. This indicates that the NN model was still able to deal with the more limited training examples of the material II's fatigue data. The necessary fatigue information and regularity contained in the training examples could be captured fairly well by the NN model to produce generally consistent fatigue life prediction results. This is also indicated by the r^2 values shown in Figures 9.14 to 9.20.

Nevertheless, the MSE values obtained were not as high as those of material I. The MSE values of the fatigue life prediction results of material II for the various training sets A, B, and C are summarized in Table 9.4. From Table 9.4, it can be seen that the MSE values were reduced to the lower values

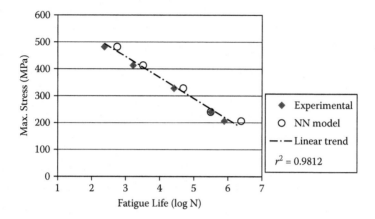

FIGURE 9.14
The *S–N* curves based on experimental data and as predicted by the NN model for $R = 0.5$.

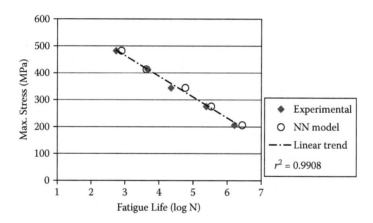

FIGURE 9.15
The *S–N* curves based on experimental data and as predicted by the NN model for $R = 0.7$.

than those of material I. This may happen because of the lesser examples used as the training data set for the NN model. Also, it can be observed that there is a variation in the MSE values of the NN prediction as the training sets are varied from T–T to C–C condition mode. The variation in the MSE values ranges between 2% and 6%. On average, the MSE measured for material II is 0.22 ± 0.03.

Based on the MSE values measured, better prediction results were achieved when using training set of R values with far-separated position in the CLD region, namely, training sets B and C. The position of the stress ratios in the two training sets allows the wider spread of fatigue data than that of the stress ratios in training set A. However, training set C, with R values from two diametrically opposed sectors of the CLD ($R = 0.1$, T–T sector; $R = 10$,

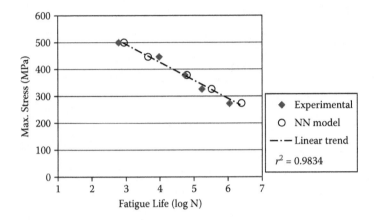

FIGURE 9.16
The *S–N* curves based on experimental data and as predicted by the NN model for $R = 0.8$.

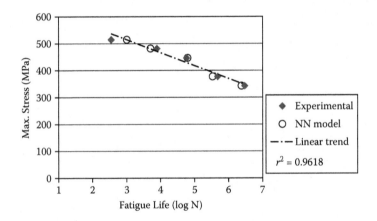

FIGURE 9.17
The *S–N* curves based on experimental data and as predicted by the NN model for $R = 0.9$.

C–C sector), gave better prediction with the lowest MSE value of 0.19 compared with training set B. This may be explained as the training set consists of R values with symmetrical position in the CLD region. The strategic position of the stress ratios may have provided the best distribution of the fatigue information that in turn resulted in the best prediction results among the three training sets.

Furthermore, it is important to point out that for material II, more R values are involved than those of material I, which means a wider spectrum of fatigue states was handled. This clearly indicates the ability of the NN model to deal with a wide spectrum of variable amplitude fatigue loading, which illustrates another value of the current study.

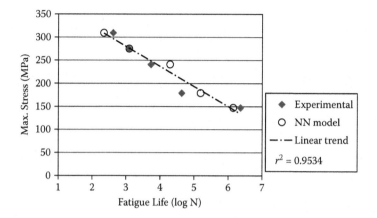

FIGURE 9.18
The *S–N* curves based on experimental data and as predicted by the NN model for *R* = −1.

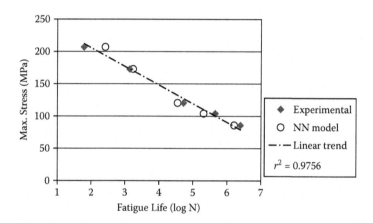

FIGURE 9.19
The *S–N* curves based on experimental data and as predicted by the NN model for *R* = −2.

The corresponding *S–N* curves of the prediction results based on training set C for material II are presented in Figures 9.14 to 9.20.

As previously mentioned, the largest discrepancies in the NN fatigue life prediction compared with that of experimental data occurred at stress ratio *R* = −0.5, as shown by Figure 9.20. It should be pointed out that the worst prediction in fatigue life occurred at stress level between 200 and 300 MPa. At another stress level, the fatigue life prediction was consistent with that of the experimental values. On the whole, a significant value of the coefficient of determination r^2 = 0.9081 was still achievable at this stress ratio. This means that the NN model still yields reasonably accurate fatigue life prediction results at a stress ratio of −0.5.

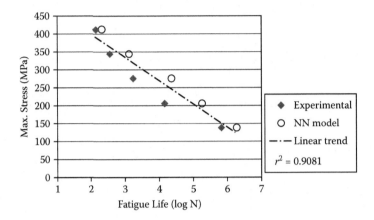

FIGURE 9.20
The *S–N* curves based on experimental data and as predicted by the NN model for $R = -0.5$.

TABLE 9.4

The MSE of the Fatigue Life Prediction Results of Material II for Various Training Sets

Training Set	R Values	The MSE Prediction Value
Set A	0.1 (T–T) and 0.5 (T–T)	0.25
Set B	0.1 (T–T) and –1 (T–C)	0.23
Set C	0.1 (T–T) and 10 (C–C)	0.19

Referring to the r^2 values calculated in Figures 9.14 to 9.20, the accuracy of the fatigue life predictions of the composite can be ascertained. The highest r^2 value of 0.9908 was achieved at $R = 0.7$, while the lowest value of 0.9081 was accomplished at $R = -0.5$. These findings proved that the use of 10 fatigue data as training examples, which is 22% of the total fatigue data, is sufficient to make a reliable fatigue life prediction of the composite.

Another significant finding of the results is that the fatigue lives predicted at a lower level of maximum stress ($\sigma_{max} < 150$ MPa) match closely the corresponding fatigue lives obtained from the experiment. The accuracy in low stress fatigue life prediction is useful in the context of designing components or structures to their fatigue limit or endurance limit. The endurance limit represents the largest value of stress that will not cause failure for an infinite number of cycles or infinite life. This information is of paramount importance to the design engineers.

9.5.2 Sensitivity Analysis

In the current study, the fatigue life predictions using the NN model have used 10 hidden nodes. The number was randomly picked from a range of

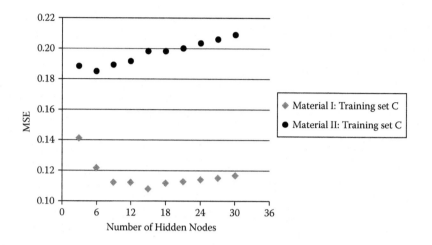

FIGURE 9.21
The sensitivity of the number of hidden nodes on the MSE values.

2–30 hidden nodes. Therefore, it is interesting to observe the effect of the number of hidden nodes on the fatigue life prediction results in term of the MSE values, as demonstrated in Figure 9.21.

It can be clearly seen that the number of hidden nodes of 10 was not the optimum number of hidden nodes with the smallest value of MSE. From Figure 9.21, the optimum number of hidden nodes for materials I and II was 15 and 6, respectively.

In addition, it can be observed that for material II, which used less fatigue data examples as the training set, the MSE value of the NN model prediction tends to increase or the quality of the NN model prediction reduced as the hidden node number increased. This can be clearly observed in the range of hidden node number between 6 and 30. No improvement in the NN prediction quality with the increase of hidden node number indicates that there is no significant "fatigue information" could be further extracted from the training set by the NN model. This could be easily understood by considering that, in principle, NN works based on learning from the training examples available.

It is also important to note that the range of hidden node number examined in the current study is comparable with that used by previous researchers. For examples Freire et al. (2005) used a range of 8–23 hidden nodes, while a range of 8–36 hidden nodes was used by Vassilopoulos et al. (2007).

9.6   Conclusions

An efficient MLP-based NN model for polymeric composites fatigue life prediction under variable amplitude loading, which optimizes limited fatigue data available, has been developed. The NN model was developed using a Bayesian regularization scheme incorporated within the LM algorithm. The NN model predicted the fatigue life of the composites with reasonable and comparable accuracy levels. Although only two stress ratios were used in the training set, the NN model developed was able to generalize well and gave a reasonably accurate fatigue life prediction under a wide range of stress ratio values. The reliability and accuracy of the NN prediction were quantified by a small MSE value.

Using 10 hidden nodes, for materials I and II, the best prediction results were achieved using training sets of 0.1, −2 and 0.1, 10, respectively. The corresponding MSE values obtained are 0.11 and 0.19, respectively, for materials I and II. As the number of hidden nodes used in the current NN model was not the optimum, the reliability and accuracy in the fatigue life prediction may well be improved further. Sensitivity analysis carried out in a range of 2 and 30 hidden nodes revealed that the optimum number of hidden nodes for materials I and II was 15 and 6, respectively. The corresponding optimum MSE values were 0.108 and 0.185, respectively.

Moreover, it is important to note that the best prediction results were achieved when using a training set of R values with a far-separated position in the CLD region, namely a training set of 0.1 and −2 for material I and 0.1 and 10 for material II. In addition, the training set of 0.1 and 10 is considered as special, because the stress ratios contained in the training set have symmetrical positions in the CLD region. The strategic position of the stress ratios may have provided the best distribution of the fatigue information, which in turn resulted in the best prediction results among the other training sets.

Further confirmation and validation to other composite materials need to be done, such as for ceramic matrix or carbon–carbon composites. The fatigue behavior of the composites should be different with the one of polymeric matrix composites. It is interesting for further observation and study.

In addition, for material II based on the high values of the coefficient r^2 for each testing set, the use of 10 fatigue data as the training examples could be considered as sufficient for the NN model prediction. However, this still needs further confirmation and examination for other types of composites. Moreover, it will be valuable to find the sufficient training data set for various types of composites in relation to the full utilization of fatigue data available.

References

Al-Assaf, Y., and El-Kadi, H. 2001. Fatigue life prediction of unidirectional glass fiber/ epoxy composite laminate using neural networks. *Composites Structures* 53:65–71.

Aymerich, F., and Serra, M. 1998. Prediction of fatigue strength of composite laminates by means of neural networks. *Key Engineering Materials* 144:231–240.

Bishop, C. M. 1995. *Neural networks for pattern recognition.* Oxford: Oxford University Press.

DOE/MSU. 2007. *DOE/MSU composite material fatigue database.* Bozeman, MT: Montana State University.

El-Kadi, H., and Al-Assaf, Y. 2002. Prediction of the fatigue life of unidirectional glass fiber/epoxy composite laminae using different neural network paradigms. *Composites Structures* 55:239–246.

Fletcher, R. 1980. *Practical methods of optimization.* Chichester: John Wiley & Sons, Ltd.

Foresee, F. D., and Hagan, M. T. 1997. Gauss-Newton approximation to Bayesian learning. *IEEE International Conference on Neural Networks* 3:1930–1935.

Freire, R. C. S., Jr., Neto, A. D. D., and de Aquino, E. M. F. 2005. Building of constant life diagrams of fatigue using artificial neural networks. *International Journal of Fatigue* 27:746–751.

Freire, R. C. S., Jr., Neto, A. D. D., and de Aquino, E. M. F. 2007. Use of modular networks in the building of constant life diagrams. *International Journal of Fatigue* 29:389–396.

Hagan, M. T., and Demuth, H. B. 1996. *Neural network design.* Boston: PWS Publishing.

Haykin, S. 1994. *Neural networks: a comprehensive foundation.* New York: Cambridge University Press.

Lee, J. A., and Almond, D. P. 2003. A neural-network approach to fatigue life prediction. In *Fatigue in composites*, ed. B. Harris, 569–589. Cambridge: Woodhead Publishing.

Mandell, J. F., and Samborsky, D. D. 1997. *DOE/MSU composite material fatigue database: test, methods, material and analysis.* SAND97-3002. Albuquerque, NM: Sandia National Laboratories.

Nocedal, J., and Wright, S. J. 2006. *Numerical optimization.* 2nd ed. New York: Springer.

Reifsnider, K. L. 1991. *Fatigue of composite materials.* Amsterdam: Elsevier.

Vassilopoulos, A. P., Georgopoulos, E. F., and Dionysopoulos, V. 2007. Artificial neural networks in spectrum fatigue life prediction of composite materials. *International Journal of Fatigue* 29:20–29.

Zhang, Z., and Friedrich, K. 2003. Artificial neural networks applied to polymer composites: a review. *Composites Science and Technology* 63:2029–2044.

10

Free Vibration Analysis and Optimal Design of the Adhesively Bonded Composite Single Lap and Tubular Lap Joints

M. K. Apalak

CONTENTS

ABSTRACT   This study focuses on the three-dimensional free vibration analysis and modal stress analysis of adhesively bonded composite single lap and tubular lap joints subjected to clamped-free conditions, and on the effects of fiber angle, fiber volume fraction, overlap length, plate/tube thickness, and tube radius on the natural frequencies and the mode shapes of adhesive joints using the back-propagation artificial neural network (ANN) method. Free vibration and stress analyses were carried out using the finite element method for random values of fiber angle, fiber volume fraction, overlap length, plate/tube thickness, and tube inner radius, so that the proposed ANN models can be trained successfully to predict the natural frequencies as well as the corresponding modal strain energies. The ANN models indicated that fiber angle was a more dominant parameter than fiber volume fraction on the natural frequencies and corresponding mode shapes, and plate/tube thickness, tube inner radius, and overlap length were important geometrical design parameters, whereas the adhesive thickness had a minor effect. In addition, genetic algorithm was combined with the trained ANN models to achieve the optimal joint design that satisfies maximum natural frequency and minimum modal strain energy conditions for each of the natural frequencies and mode shapes of the adhesively bonded composite single lap and tubular lap joints.

10.1   Introduction

The adhesive bonding technique is used succesfully for joining composite materials since it removes difficulties in the joining process (e.g., stiffness losses due to holes, unsuitable load transfer via bolt-hole contact) and provides uniform load distribution between adherends (Adams and Wake, 1984; Kinloch, 1987; Tong and Steven, 1999). Because adhesive joints serve under harmonic or impulsive loads as well as static loads, they experience harmonic or damped vibrations. These periodic loads cause local fatigue in stress concentration regions in both adhesive layer and adherends. It is essential that the dynamic response of adhesive joints to the dynamic loads be understood to achieve a long service life for an adhesive joint design. The free vibration analysis provides the natural frequencies and mode shapes of adhesive joints; therefore, the adhesive joint can be designed depending on its most critical mode shape and its service life, and its dynamic strength can be also improved. However, the effects of the mechanical properties and geometrical parameters of adhesive joints on their dynamic behavior are important in achieving a high-performance adhesive joint design.

The free or damped transient vibrations of the structural members, such as stiffened composite or metal beams, plates, and shells, in which the adhesive bonding method is widely used, have been investigated in detail (Reddy, 1997; Yuceoglu et al., 1996; Yuceoglu and Ozerciyes, 2000, 2003). Geometrical parameters, such as overlap length and adherend thickness, play an important role on the natural frequencies and logarithmic decrement of metal or composite adhesive single lap joints (Saito and Tani, 1984; Ko et al., 1995). The fundamental frequencies of patched composite plates are affected by the patched size and adhesive modulus; thus, increasing the patch size of a patched composite plate results in a stiffer plate with higher frequencies, and the natural frequencies of a patched plate with lower adhesive modulus become lower than those of an unpatched plate (Lin and Ko, 1997). Adhesive modulus and Poisson's ratio have a negligible effect on the transverse natural frequencies of the cantilevered adhesive single lap joint, but adhesive stiffness affects even mode shapes. Thus, a flexible adhesive layer markedly affects the mode shapes more than a stiffer adhesive layer does (He and Oyadiji, 2001). Consequently, the higher stresses occur in a stiffer adhesive layer, which probably suffers from fatigue failure and debonding more than the softer adhesive layer. In addition, the resonant amplitudes of adhesive single lap joints can be significantly reduced with increased damping (Kaya et al., 2004).

The adhesively bonded multilayered composite beams, plates, and shells in engineering structures undergo dynamic loads, and their fatigue life is dependent on the vibration characteristics. The free, flexural vibrations of relatively thick, orthotropic, composite base plate or panel reinforced by a stiffening plate strip located in any region of the composite plate were investigated in detail, and the material characteristics of the adhesive layer were found to have significant effects on the deformations of bonded multilayer composite plates and on their natural frequencies and mode shapes (Yuceoglu et al., 1996; Yuceoglu and Ozerciyes, 2000, 2003, 2005). In addition, the influence of structural defects on the frequencies of composite beams bonded by an adhesive layer including different volumes of defects differs from the prevailing opinion about the effect of structural defects on modal parameters, and a different damping mechanism was reported (Yang et al., 1998).

The tubular and canonical multilayered composite structures are also widely used in engineering designs subjected to dynamic external loads. Rubber layers are embedded into a graphite composite plate in order to reduce vibrations. These rubber layers considerably affect modal parameters, such as natural frequencies, damping, and mode shapes, and significantly increase structural modal damping and decrease natural frequencies (Taleghani and Pappa, 1996). The dynamic response of an adhesive tubular lap joint with viscoelastic adhesive layer under a harmonic axial load is sensitive to the adhesive loss factor and is slightly affected by a central annular void in the bond area with a size less than 40% of the overlap length in adhesives with a larger loss factor (Vaziri and Nayeb-Hashemi, 2002). The vibration characteristics of functionally graded cylindrical shells are similar to those of isotropic cylindrical shells, and the natural frequencies of these shells are dependent on the constituent volume fractions and boundary conditions (Loy et al., 1999; Pradhana et al., 2000).

The free edges of the adhesive layer of the lap joints undergo considerable stress and strain, being subjected to tensile and bending loads (Adams and Wake, 1984; Kinloch, 1987; Tong and Steven, 1999). The free edges of the adhesive-adherend interfaces experience peak stresses, whereas negligible stress levels appear in the middle of the overlap region. The presence of adhesive fillets around the adhesive free ends results in stress concentrations occurring around the adherend corners inside the adhesive fillets at the free edges of the adhesive-adherend interfaces. A probable crack or local damage initiates here and propagates along the adherend-adhesive interfaces and toward the free surface of the adhesive fillet. The elastic strain energy reaches a maximal level at these critical adhesive regions, and the first local failure happens when the adhesive elastic strain energy reaches a specific level. The modified von Mises criterion is widely used for predicting adhesive failure. To reduce these peak adhesive stresses, the adherend free edges are modified by tapering or scarfing to a knife edge or the material of the adherends is degraded through the adherend thickness. Either increasing the overlap length or decreasing the adhesive thickness can also reduce peak adhesive stresses (Adams and Wake, 1984; Kinloch, 1987; Tong and Steven, 1999). In case of impact, periodic external loads, the peak stress and critical strain energy around the free edges of the adhesive-adherend interfaces become critical depending on the deformed shape of the excited adhesive joint. An adhesive joint is designed such that peak stress concentrations in the adhesive layer are reduced as much as possible, and its first natural frequency is increased to avoid the coincidence of the natural frequency with the frequency of the applied external load. This yields an optimization problem. The modal strain energy index is important as well as the maximization of the natural frequencies. Therefore, natural frequency and mode shape are the main parameters (Jaishi and Ren, 2005). The modal strain energy can be used successfully for the identification of structural behavior and location of structural damage since modal strain energy is an efficient damage indicator and an algorithm that uses modal strain energy and is more sensitive to local damage than other mode shape-based functions, whereas natural frequencies provide global information and have a favorable effect on the conditions of the optimization problem (Lim and Kashangaki, 1994; Doebling et al., 1997; Stubbs and Kim, 1996; Shi and Law, 1998; Shi et al., 2000; Jaishi and Ren, 2007). Consequently, modal strain energy and natural frequencies are widely used as independent objective functions.

One- or two-dimensional free vibration analyses of adhesively bonded composite structures yield important information about the effects of the mechanical properties of adhesive layers as well as the geometrical parameters of the adhesive joint, but neglects the free vibration behavior of the composite structure along the third dimension. This study investigates the three-dimensional free vibration of the adhesively bonded composite single lap and tubular lap joints using the finite element method and the effects of the design parameters of the adhesive and composite plates/tubes (fiber angle, fiber volume fraction, adhesive thickness, overlap length and plate/tube thickness, and tube radius) using the proposed artificial neural networks (ANNs), and an optimal design for each of the adhesively bonded composite single lap and tubular lap joints was determined by combining genetic algorithm (GA) and ANN models based on the first 10 natural frequencies and corresponding modal strain energies of both adhesive joints.

10.2 Dynamic Equations of Motion

The dynamic equations of motion of a structure can be derived by using either Lagrange equations and Hamilton's principle (Rao, 1989; Cook, 1981). The Lagrange equations are given by

$$\frac{d}{dt}\left\{\frac{\partial L}{\partial \dot{q}}\right\} - \left\{\frac{\partial L}{\partial q}\right\} + \left\{\frac{\partial R}{\partial \dot{q}}\right\} = \{0\} \tag{10.1}$$

where

$$L = T - \pi_p \tag{10.2}$$

is called the Lagrange function, T is the kinetic energy, π_p is the potential energy, R is the dissipation function, q is the nodal displacement, and $\dot{q}$ is the nodal velocity. The kinetic and potential energies of each element of the structure that was divided into E elements can be written as

$$T^{(c)} = \frac{1}{2}\iiint_{V^{(e)}} \rho \dot{u}^T \dot{u}\, dV \tag{10.3}$$

$$\pi_p^{(e)} = \frac{1}{2}\iiint_{V^{(e)}} \varepsilon^T \sigma\, dV - \iint_{S_1^{(e)}} u^T \phi_1\, dS_1 - \iiint_{V^{(e)}} u^T \phi_2\, dV \tag{10.4}$$

where ρ is density, and u and $\dot{u}$ are the displacement and velocity vectors, respectively. The dissipation function of each element

$$R^{(e)} = \frac{1}{2}\iiint_{V^{(e)}} \mu \dot{u}^T \dot{u}\, dV \tag{10.5}$$

where μ is the damping coefficient. Using the finite element method, the overall system equations of motion can be written as

$$[M]\ddot{Q}(t) + [C]\dot{Q}(t) + [K]Q(t) = P(t) \tag{10.6}$$

where $\ddot{Q}(t)$ is the vector of nodal accelerations, and

$$\text{mass matrix } [M] = \sum_{e=1}^{E} \iiint_{V^{(e)}} \rho [N]^T [N] dV \tag{10.7}$$

$$\text{element stiffness matrix } [K] = \sum_{e=1}^{E} \iiint_{V^{(e)}} [B]^T [D][B] dV \tag{10.8}$$

$$\text{element damping matrix } [C] = \sum_{e=1}^{E} \iiint_{V^{(e)}} \mu [N]^T [N] dV \tag{10.9}$$

$$\text{total load vector } [P] = \sum_{e=1}^{E} \iiint_{S_1^{(e)}} [N]^T \{\phi_1\} dS_1 + \iiint_{V^{(e)}} [N]^T \{\phi_2\} dV \tag{10.10}$$

where $[N]$ is the matrix of shape functions, $[B]$ is the matrix of coordinate derivatives of shape functions, $[D]$ is the matrix of elastic constants, $\{\phi_1\}$ is the surface traction vector, and $\{\phi_2\}$ is the body force vector. The equation of motion for an undamped system without external force becomes

$$[M]\ddot{Q}(t) + [K]Q(t) = 0 \tag{10.11}$$

The solution of this eigenvalue problem

$$\det\left([K] - \lambda^2[M]\right) = \{0\} \tag{10.12}$$

Each eigenvalue λ_i is associated with an eigenvector Q_i, which is called a natural mode. The @@@block Lanczos eigenvalue extraction method was used for the calculation of eigenvalues and eigenvectors since the models have large degrees of freedom (Rao, 1989; Cook, 1981; Grimes et al., 1994).

10.3 Micromechanics of Composite Materials

The laminated composite plates and tubes are made of IM6 (graphite fiber) and 3501-6 (epoxy system). The mechanical and physical properties of the

TABLE 10.1

Mechanical Properties of Graphite Fiber (IM6), Epoxy (3501-6), and Lamina
(IM6/3501-6) for $V_f = 63.5\%$ (Apalak and Yildirim, 2009)

Property	Unit	Fiber	Epoxy	Lamina
ρ	kg/m³	1743.834	1264.972	1552.289
E_{11}	GPa	259.105	4.344	157.218
E_{22}, E_{33}	GPa	13.927	4.344	9.309
G_{12}, G_{13}	GPa	50.952	1.597	5.723
G_{23}	GPa	8.274	1.597	3.475
v_{12}, v_{13}		0.26	0.36	0.3
v_{23}		0.33	0.36	0.34

graphite fiber and epoxy adhesive are given in Table 10.1. The engineering
constants of the unidirectional lamina (IM6/3501-6) can be calculated by
using the micromechanical approach based on the engineering constants of
IM6 fiber and 3501-6 epoxy in the material coordinates (x_1, x_2, x_3) as

$$E_1 = E_f V_f + E_m V_m \tag{10.13}$$

$$E_2 = \frac{E_f E_m}{E_f V_m + E_m V_f} \tag{10.14}$$

$$G_{12} = \frac{G_f G_m}{G_f V_m + G_m V_f} \tag{10.15}$$

$$V_{12} = v_f V_f + v_m V_m \tag{10.16}$$

$$\rho = \rho_f V_f + \rho_m V_m \tag{10.17}$$

$$\frac{v_{21}}{E_2} = \frac{v_{12}}{E_1} \tag{10.18}$$

where E_1 is the longitudinal elasticity modulus, E_2 is the transverse elasticity
modulus, v_{12} and v_{21} are the major and minor Poisson's ratios, respectively,

G_{12} is the shear modulus, and ρ is the density (Reddy, 1997). The elastic constants in the material coordinates are

$$[Q] = \begin{bmatrix} Q_{11} & Q_{12} & 0 \\ Q_{21} & Q_{22} & 0 \\ 0 & 0 & Q_{66} \end{bmatrix} \tag{10.19}$$

where

$$Q_{11} = \frac{E_1}{1 - v_{12}v_{21}}$$

$$Q_{12} = \frac{v_{12}E_1}{1 - v_{12}v_{21}}$$

$$Q_{21} = \frac{v_{21}E_1}{1 - v_{12}v_{21}} = Q_{12} \tag{10.20}$$

$$Q_{22} = \frac{E_2}{1 - v_{12}v_{21}}$$

$$Q_{66} = G_{12}$$

and in the problem coordinate system (x,y,z) as

$$[\bar{Q}] = [T]^{-1}[Q][T] \tag{10.21}$$

where the transformation matrix is

$$[T] = \begin{bmatrix} n^2 & m^2 & 2mn \\ m^2 & n^2 & -2mn \\ -mn & mn & (n^2m^2) \end{bmatrix} \tag{10.22}$$

with = $\sin(\alpha)$ and $n = \cos(\alpha)$, and the fiber orientation angle α.

10.4   Free Vibration Analysis of Composite Single Lap Joint

The adhesively bonded composite single lap joint is formed by bonding two unidirectional composite (adherends) plates via an epoxy adhesive layer, as

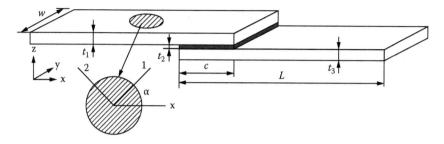

FIGURE 10.1
Clamped-free adhesively bonded composite single lap joint (Apalak and Yildirim, 2009).

shown in Figure 10.1. The adhesive joint is clamped through the back surface of the upper plate. The adhesive joint has plate thickness $t_1 = t_3 = 0.5–2$
mm, plate length $L = 160$ mm, overlap length $c = 10–50$ mm, joint width W
$= 25$ mm, and adhesive thickness $t_2 = 0.1–0.5$ mm. The laminated composite plates are made of IM6 (graphite fiber) and 3501-6 (epoxy system). A ply
(IM6/3501-6) has a thickness of 0.1335 mm and a fiber volume fraction of
0.635 (Table 10.1).

The first six natural frequencies were extracted and the corresponding
mode shapes were calculated using the ABAQUS/Standard Finite Element
Software ABAQUS (ABAQUS). The plates were modeled with a four-node
shell finite element S4R with three integration points throughout the element
thickness and the adhesive layer with an eight-node solid element C3D8R
(Figure 10.2). First, the free vibration analysis was carried out for an adhesively bonded composite single lap joint with plate thickness $t_1 = t_3 = 1.0$ mm,
overlap length $c = 20$ mm, and adhesive thickness $t_2 = 0.2$ mm. Each plate
is made of eight plies with a fiber angle $\alpha = 0°$ and a fiber volume fraction
$V_f = 0.635$. The first six natural frequencies appear as transverse (bending)
and torsional mode shapes (Figure 10.3). The natural frequencies of bending
modes are $\omega_1 = 16.7$, $\omega_2 = 101.5$, and $\omega_4 = 295.0$ Hz. The torsional modes appear
at natural frequencies $\omega_3 = 104.8$, $\omega_5 = 319.5$, and $\omega_6 = 417.9$ Hz.

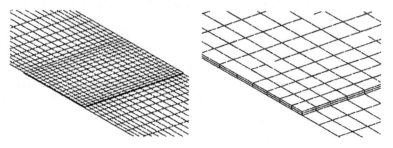

FIGURE 10.2
Finite element model of the overlap region of adhesively bonded composite single lap joint
(Apalak and Yildirim, 2009).

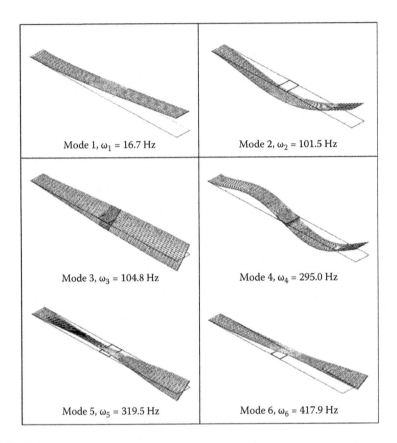

FIGURE 10.3
The first six natural frequencies and corresponding mode shapes of the clamped-free adhesively bonded composite single lap joint ($t_2 = 0.2$ mm, $t_1 = t_3 = 1.0$ mm, $L = 160$ mm, $c = 20$ mm, $W = 25$ mm, $\alpha = 0°$, $V_f = 0.635$) (Apalak and Yildirim, 2009).

The mechanical and physical properties of the epoxy adhesive, such as the modulus of elasticity, Poisson's ratio, and density, have negligible effects on the natural frequencies and modal strain energies of the adhesive single joint (Yildirim, 2006). Consequently, the mechanical and physical properties of the adhesive layer and the adhesive thickness, $t_2 = 0.2$ mm, were kept constant throughout the study.

Table 10.2 shows the effect of the fiber angle (α) between 0° and 90° on the first six normalized natural frequencies and corresponding modal strain energies. Increasing the fiber angle results in a decrease by 51–74% in the natural frequencies and by 45–93% in the corresponding modal strain energies. Therefore, fiber angle α appears as an effective design parameter. However, a good joint design requires that natural frequency be maximized and modal strain energy be minimized. As the fiber volume fraction V_f varies from 40% to 70%, the first six natural frequencies increased by 20–24% (Table 10.3),

TABLE 10.2

Effect of Fiber Angle (α) on (a) the First Six Natural Frequencies and (b) the Corresponding Modal Strain Energies of the Adhesively Bonded Composite Single Lap Joint (Apalak and Yildirim, 2009)

(a)

Mode	Normalized Natural Frequencies (ω_i/ω_{max})					
	$\bar{\omega}_1$	$\bar{\omega}_2$	$\bar{\omega}_3$	$\bar{\omega}_4$	$\bar{\omega}_5$	$\bar{\omega}_6$
α	16.7	101.5	172.2	295.0	319.4	417.9
0°	1.0000	1.0000	0.6084	1.0000	1.0000	1.0000
15	0.5915	0.5897	1.0000	0.5949	0.7764	0.8254
30	0.3626	0.3658	0.6267	0.5154	0.6398	0.5258
45	0.2808	0.2832	0.4819	0.4068	0.5074	0.4128
60	0.2553	0.2571	0.4371	0.3779	0.4110	0.3491
75	0.2554	0.2569	0.4375	0.3576	0.3539	0.3476
90	0.2598	0.2614	0.4453	0.3278	0.3606	0.3535

(b)

Mode	Normalized Modal Strain Energies					
α	1	2	3	4	5	6
0°	15.6	541.1	1210.4	4628.0	3449.9	10,554.6
0°	1.0000	1.0000	0.3257	1.0000	0.6788	1.0000
15	0.2943	0.2058	1.0000	0.1275	1.0000	0.2838
30	0.1181	0.1012	0.3232	0.2791	0.2408	0.1326
45	0.0749	0.0712	0.2420	0.1751	0.3057	0.0992
60	0.0645	0.0649	0.2342	0.1520	0.1671	0.0862
75	0.0649	0.0638	0.2435	0.0923	0.2121	0.1068
90	0.0676	0.0679	0.2578	0.0796	0.2187	0.1111

TABLE 10.3

Effect of Fiber Volume Fraction (V_f) on (a) the First Six Natural Frequencies and (b) the Corresponding Modal Strain Energies of the Adhesively Bonded Composite Single Lap Combined (Apalak And Yildirim, 2009)

(a)

	Normalized Natural Frequencies (ω_i/ω_{max})					
Mode	$\bar{\omega}_1$	$\bar{\omega}_2$	$\bar{\omega}_3$	$\bar{\omega}_4$	$\bar{\omega}_5$	$\bar{\omega}_6$
V_f (%)	17.8	108.4	117.4	315.1	356.8	449.0
40.0	0.7992	0.7987	0.7638	0.7993	0.7653	0.7948
45.5	0.8423	0.8417	0.7935	0.8423	0.7956	0.8359
48.2	0.8625	0.8619	0.8097	0.8626	0.8121	0.8556
50.9	0.8819	0.8814	0.8271	0.8819	0.8295	0.8747
56.4	0.9186	0.9182	0.8659	0.9186	0.8683	0.9117
59.1	0.9360	0.9356	0.8878	0.9360	0.8899	0.9297
61.8	0.9528	0.9525	0.9117	0.9528	0.9136	0.9475
67.3	0.9848	0.9847	0.9673	0.9848	0.9681	0.9826
70.0	1.0000	1.0000	1.0000	1.0000	1.0000	1.0000

(b)

	Normalized Modal Strain Energies					
Mode	1	2	3	4	5	6
V_f (%)	18.4	633.0	514.6	5424.7	3030.0	12,494.8
40.0	0.5818	0.5805	0.5281	0.5815	0.5243	0.5781
45.5	0.6577	0.6565	0.5785	0.6575	0.5743	0.6518
48.2	0.6957	0.6945	0.6073	0.6955	0.6029	0.6889
50.9	0.7337	0.7325	0.6389	0.7335	0.6346	0.7263
56.4	0.8097	0.8088	0.7129	0.8096	0.7087	0.8019
59.1	0.8477	0.8469	0.7565	0.8477	0.7525	0.8404
61.8	0.8858	0.8851	0.8056	0.8857	0.8021	0.8793
67.3	0.9619	0.9617	0.9256	0.9619	0.9239	0.9589
70.0	1.0000	1.0000	1.0000	1.0000	1.0000	1.0000

whereas the modal strain energies increased by 42–48%. Fiber volume fraction V_f is also another design parameter.

For a fiber angle $\alpha = 0°$ and a fiber volume fraction $V_f = 0.635$, increasing the overlap length c by between 10 and 50 mm results in increases by 17–26% in natural frequencies and increases by 10–60% in modal strain energies (Table 10.4). The natural frequencies and modal strain energies increase because the adhesive joint becomes stiffer as the overlap length is increased. Consequently, the overlap length is an important geometrical design parameter for an optimal joint design.

Table 10.5 shows the effect of plate thickness $t_1 = t_3$, which is varied between 0.53 and 2.0 mm, on natural frequencies and modal strain energies. Plate thickness is a geometrical parameter considerably affecting natural frequencies and modal strain energies. Thus, increasing the plate thickness results in increases of 52–73% in natural frequencies and increases of 85–100% in modal strain energies.

It is evident that natural frequencies and modal strain energies are strongly dependent on fiber angle and volume fraction, as well as geometrical parameters such as overlap length and plate thickness. An optimal design of the adhesively bonded composite single lap joint can be found within the specified limits of the design parameters, except for the mechanical properties of the adhesive layer and the adhesive thickness becasue of their negligible effects. An optimal joint design can be directly searched among the 1,771,200 models with various design variables, such as a fiber angle $0,1 \leq \alpha \leq 90°$, a fiber volume fraction $0.40,0.41 \leq V_f \leq 0.70$, a support length $10,11 \leq c \leq 50$ mm, and a plate thickness $0.5,0.6 \leq t_1 \leq 2.0$ mm. Based on a three-dimensional free vibration analysis of each joint model, the first six eigenvalues of each model are extracted, and the mode shape for each natural frequency and the subsequent stress and strain distributions are determined for the calculation of the modal strain energy. Therefore, the direct search for an optimal joint design is a challenge. ANNs offer a more convenient means of predicting the first six natural frequencies and modal strain energies for various values of the design parameters (Simpson, 1990; Hagan and Menhaj, 1994). For this purpose, it is necessary to design an appropriate ANN model and to train it based on the training data generated from the free vibration analysis using the finite element method for random values of the design parameters. Later, the performance of the proposed neural network model can be tested with the testing data generated in a similar manner. The neural network models can be trained and tested with minimum data in comparison with the direct search. Consequently, the couple effects of these design parameters on natural frequencies and modal strain energies can be easily investigated, and an optimal joint design can be searched with minimum calculation using GA (Holland, 1975; Michalewicz, 1999; Goldberg, 1989, 2002) combined with these trained ANNs.

To predict the first six natural frequencies and the corresponding modal strain energies using a proposed neural network model, the training and

TABLE 10.4

Effect Of Overlap Length (c) on (a) the First Six Natural Frequencies and (b) the Corresponding Modal Strain Energies of the Adhesively Bonded Composite Single Lap Joint (Apalak and Yildirim, 2009)

(a)

Mode	Normalized Natural Frequencies (ω_i/ω_{max})					
	$\overline{\omega}_1$	$\overline{\omega}_2$	$\overline{\omega}_3$	$\overline{\omega}_4$	$\overline{\omega}_5$	$\overline{\omega}_6$
c (mm)	20.2	116.4	124.7	354.7	357.9	497.4
10.0	0.7781	0.8271	0.8128	0.7799	0.8707	0.7946
13.6	0.7951	0.8423	0.8229	0.7983	0.8783	0.8109
20.9	0.8309	0.8763	0.8432	0.8366	0.8949	0.8446
24.6	0.8497	0.8953	0.8535	0.8565	0.9041	0.8621
28.2	0.8690	0.9157	0.8639	0.8768	0.9139	0.8801
35.5	0.9098	0.9484	0.8973	0.9189	0.9361	0.9178
42.7	0.9533	0.9733	0.9454	0.9629	0.9616	0.9577
50.0	1.0000	1.0000	1.0000	1.0000	1.0000	1.0000

(b)

Mode	Normalized Modal Strain Energies					
	1	2	3	4	5	6
c (mm)	22.7	570.4	682.2	5717.2	6106.3	15,086.1
10.0	0.6141	0.8974	0.5299	0.7387	0.4142	0.6273
13.6	0.6399	0.9148	0.5470	0.7635	0.4016	0.6524
20.9	0.6959	0.9539	0.5826	0.8163	0.3813	0.7068
24.6	0.7265	0.9760	0.6013	0.8444	0.3731	0.7363
28.2	0.7589	1.0000	0.6208	0.8735	0.3660	0.7677
35.5	0.8296	0.7923	0.8817	0.9349	0.3549	0.8360
42.7	0.9095	0.8469	0.9356	1.0000	0.3476	0.9130
50.0	1.0000	0.9070	1.0000	1.0000	1.0000	1.0000

TABLE 10.5

Effect of Plate Thickness (t_1) on (a) the First 10 Natural Frequencies and (b) the Corresponding Modal Strain Energies of the Adhesively Bonded Composite Single Lap Joint (Apalak and Yildirim, 2009)

(a)

	Normalized Natural Frequencies (ω_i/ω_{max})					
Mode	$\bar{\omega}_1$	$\bar{\omega}_2$	$\bar{\omega}_3$	$\bar{\omega}_4$	$\bar{\omega}_5$	$\bar{\omega}_6$
t_1 (mm)	35.9	216.9	220.8	420.3	632.2	675.3
0.53	0.2667	0.2694	0.2751	0.4038	0.2917	0.4837
0.80	0.4003	0.4036	0.4092	0.6053	0.4352	0.6209
0.93	0.4671	0.4705	0.4759	0.7059	0.5065	0.6221
1.07	0.5338	0.5373	0.5425	0.8064	0.5771	0.6233
1.20	0.6005	0.6039	0.6088	0.9067	0.6433	0.6281
1.47	0.7338	0.7368	0.7406	0.9984	0.7358	0.7412
1.60	0.8004	0.8029	0.8059	0.9992	0.8021	0.8063
1.73	0.8669	0.8688	0.8709	0.9996	0.8682	0.8712
2.00	1.0000	1.0000	1.0000	1.0000	1.0000	1.0000

(b)

	Normalized Modal Strain Energies					
Mode	1	2	3	4	5	6
t_1 (mm)	154.2	5330.3	3737.4	22,609.8	45,779.7	22,624.5
0.53	0.0193	0.0189	0.0207	0.0385	0.0092	0.1439
0.80	0.0647	0.0642	0.0676	0.1299	0.0318	0.4022
0.93	0.1025	0.1021	0.1063	0.2064	0.0509	0.4713
1.07	0.1528	0.1524	0.1575	0.3081	0.0769	0.5473
1.20	0.2173	0.2170	0.2228	0.4386	0.1395	0.8049
1.47	0.3959	0.3959	0.4022	0.7369	0.3952	0.4011
1.60	0.5134	0.5136	0.5194	0.8013	0.5129	0.5172
1.73	0.6522	0.6524	0.6571	0.8673	0.6518	0.6550
2.00	1.0000	1.0000	1.0000	1.0000	1.0000	1.0000

testing data were generated based on the free vibration analysis of the composite adhesive single lap joint using the finite element method for random values of the design variables, such as a fiber angle $0 \leq \alpha \leq 90°$, a fiber volume fraction $0.40 \leq V_f \leq 0.70$, a support length $10 \leq c \leq 50$ mm, and a plate thickness $0.5 \leq t_1 = t_3 \leq 2.0$ mm. The design variables, the first six natural frequencies, and the modal strain energies were collected and then used as input data for the training session of the neural network. Two training sessions were carried out for each natural frequency and modal strain energy. For the training sessions, the input for neural network models comprised fiber volume fraction, fiber angle, overlap length, and plate thickness (Figure 10.4), and the output of the neural network model consisted of the first six natural frequencies and the corresponding strain energies.

The Levenberg-Marquardt back-propagation learning algorithm (Hagan and Menhaj, 1994) was implemented. The proposed ANN model is a feedforward, two–hidden layer network. Twenty and 12 neurons were used in the first and second hidden layers, respectively. The logarithmic sigmoid transfer function (Figure 10.5a)

$$f(x) = \frac{2}{1+e^{-x}} \tag{10.23}$$

and the tangent sigmoid transfer function (Figure 10.5b)

$$f(x) = \frac{1-e^{-2x}}{1+e^{-2x}} \tag{10.24}$$

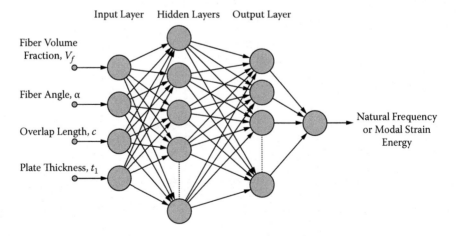

FIGURE 10.4
Proposed artifical neural network model (Apalak and Yildirim, 2009).

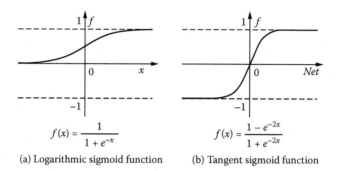

$$f(x) = \frac{1}{1 + e^{-x}}$$

(a) Logarithmic sigmoid function

$$f(x) = \frac{1 - e^{-2x}}{1 + e^{-2x}}$$

(b) Tangent sigmoid function

FIGURE 10.5
Activation functions used in the ANN model (Apalak and Yildirim, 2009).

are activation functions used throughout the hidden layers and the output layer. The training and testing data were normalized to a range of 0–1. The neural network toolbox of MATLAB (MATLAB) was used to develop a proposed neural network model. A data set is composed of 4000 patterns based on the free vibration analysis of the adhesively bonded composite single lap joint for random values of the design variables. The training phase is performed with (an input data) 3200 patterns, and the performance of ANN models was tested with (testing data) the remaining 800 patterns randomly selected from 4000 patterns. All ANN models have a layer configuration of 4, 20, 12, and 1 neurons in each layer (1–4) (Figure 10.4). For the training phase, the performance was evaluated with the mean square error between the values predicted by the proposed neural network and the test target values. During the training phase, mean square error was reduced to an order of 10^{-8} depending on the mode shape.

The free vibration analysis of the adhesively bonded composite single lap joint showed that the bending and torsional mode shapes were dominant, and that fiber angle, fiber volume fraction, overlap length, and plate thickness had an important impact on both natural frequencies and modal strain energies, whereas increasing the adhesive thickness exhibits a minor effect (Yildirim, 2006). To determine the combined effect of both these dominant design parameters on natural frequencies and strain energies, the proposed neural network models were used. An optimal value for fiber volume fraction is 0.635, used widely for a unidirectional lamina (IM6/3501-6) considering the lamina strength. Figure 10.6 shows the combined effect of fiber angle α and overlap length c on the first six natural frequencies. The six natural frequencies decrease considerably with increasing fiber angle α, and its effect disappears when fiber angle $\alpha = 60°$, whereas the natural frequencies increase with increasing overlap length c. However, fiber angle α is a more effective design parameter on natural frequencies than overlap length c. Similar effects are observed for modal strain energies. Figure 10.7 shows the combined effect of fiber angle α

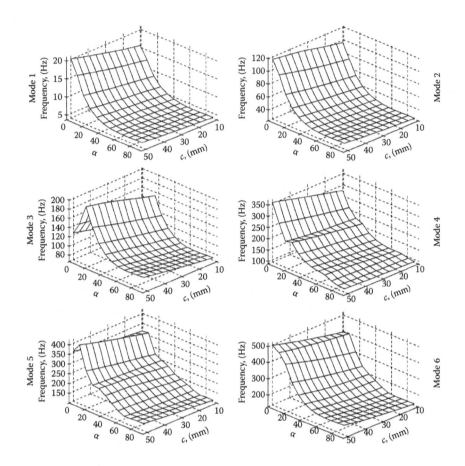

FIGURE 10.6
Combined effect of fiber angle (α) and overlap length (c) on the first six natural frequencies of the clamped-free adhesively bonded composite single lap joint ($t_1 = t_3 = 1.0$ mm, $t_2 = 0.2$ mm, V_f = 0.635) (Apalak and Yildirim, 2009).

and plate thickness t_1 on the first six natural frequencies. Increasing fiber angle α leads to a decrease in natural frequencies. The natural frequencies exhibit proportional increases to plate thickness t_1. In particular, when fiber angle $\alpha = 0°$, the effect of plate thickness t_1 becomes obvious. Thus, peak natural frequencies appear when plate thickness is maximal and α = 0°. Plate thickness t_1 is a more dominant design parameter compared to overlap length c (Figure 10.8). Peak natural frequencies occur when both plate thickness t_1 and overlap length c are maximal.

GAs have been successfully adapted to the optimization problems of engineering structures (Holland, 1975; Michalewicz, 1999; Goldberg, 1989, 2002). GAs are a part of evolutionary computing and use techniques inspired by evolutionary biology such as inheritance, mutation, natural selection, and recombination (or crossover). The optimization problem is solved with an

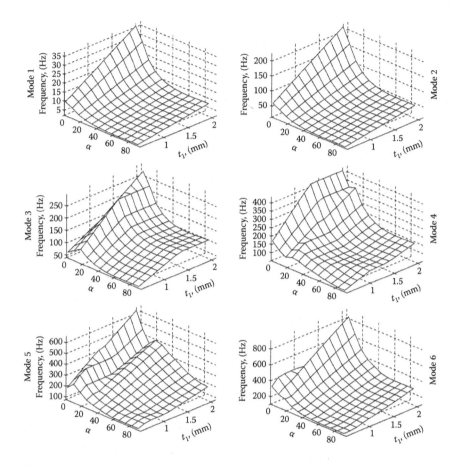

FIGURE 10.7
Combined effect of fiber angle (α) and the plate thickness (t_1) on the first six natural frequencies of the clamped-free adhesively bonded composite single lap joint ($t_2 = 0.2$ mm, $c = 20$ mm, $V_f = 0.635$) (Apalak and Yildirim, 2009).

evolutionary process that results in a best (fittest) solution. In genetic programming, a population of abstract representations (chromosomes) among the candidate solutions (individuals) to an optimization problem is evolved to achieve better solutions. In general, solutions are represented in strings composed of 0's and 1's as well as different encodings. The evolution process begins from a population of random individuals and continues for generations. In each generation, the fitness values of the population are evaluated, and multiple individuals are stochastically selected from the present population considering their fitness values, and a new population is formed by modifying (mutating or recombining) these selected multiple individuals. This population is considered as the current population of the next iteration.

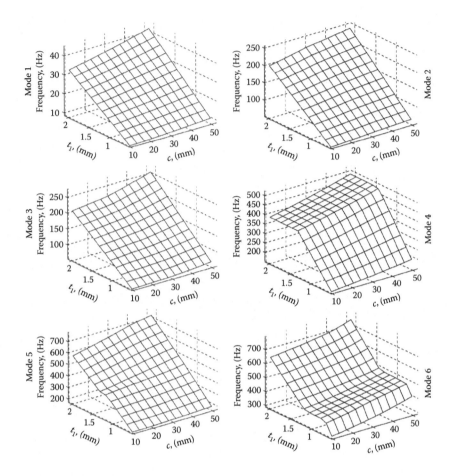

FIGURE 10.8
Combined effect of plate thickness (t_1) and overlap length (c) on the first six natural frequencies of the clamped-free adhesively bonded composite single lap joint ($t_2 = 0.2$ mm, $\alpha = 0°$, $V_f = 0.635$) (Apalak and Yildirim, 2009).

An optimal design problem of the adhesively bonded composite single lap joint requires the fundamental frequencies to be maximized and the corresponding modal strain energies to be minimized. GAs search for a most appropriate set of design parameters by evaluating an objective function in which these conditions are defined. For this purpose, either the finite element method or the proposed neural network models can be used. However, the finite element method requires a time-consuming calculation phase for each new set of design variables. The trained neural network models instantly predict all natural frequencies and corresponding modal strain energies. The GA toolbox of MATLAB (MATLAB) was used to solve the optimal joint problem defined as follows.

Find the fiber volume fraction V_f, fiber angle α, overlap length c, and plate thickness t_1 that will minimize the objective function

$$f(\omega_i, U_i) = -c_1 \times \omega_i + c_2 \times U_i \qquad (10.25)$$

where $c_1 = c_2 = 0.0–1.0$, ω_i is ith natural frequency, and U_i is the corresponding modal strain energy, subject to the following constraints: $0.40 \leq V_f \leq 0.70$, $0 \leq \alpha \leq 90°$, $10 \leq c \leq 50$ mm, and $0.5 \leq t_1 \leq 2.0$ mm. Three objective functions were used as follows (Table 10.6): (1) natural frequency and modal strain energy were equally weighted ($c_1 = c_2 = 0.5$); (2) maximization of natural frequency only ($c_1 = 1.0$ and $c_2 = 0.0$); and (3) minimization of modal strain energy only ($c_1 = 0.0$ and $c_2 = 1.0$). Maximization of natural frequencies alone yielded the following result: overlap length c, plate thickness t_1, and fiber volume fraction V_f became maximal, and fiber angle α approached zero or became smaller than $14°$. Increasing the plate stiffness resulted to natural frequencies being maximized. Minimization of modal strain energies alone led to overlap length, plate thickness, and fiber volume fraction being minimized, whereas fiber angle is maximized. As the joint stiffness is reduced, the modal strain energy is minimized. This indicates lower local deformations in the adhesive-free edges. Two objective functions yield different optimal parameter values due to the opposite responses of natural frequency and modal strain energy to the design parameters. An objective function—natural frequency and corresponding modal strain energy are equally weighted—gives a maximal support length, a fiber angle between $6°$ and $34°$, a fiber volume fraction between 40% and 70%, and a minimal plate thickness.

10.5 Composite Tubular Single Lap Joint

The clamped-free composite tubular single lap joint consists of two bonded unidirectional composite tubes (Figure 10.9), with inner and outer tube thicknesses $t_1 = t_2 = 0.5–2$ mm, tube length $L = 250$ mm, overlap length $c = 10–50$ mm, inner tube radius $R = 12.5–30$ mm, and adhesive thickness $t_3 = 0.2$ mm. The tubes are made of plies (IM6/3501-6) with a thickness of 0.1335 mm and a fiber volume fraction $V_f = 0.635$ (Table 10.1). The first 10 natural frequencies and the corresponding mode shapes of the tubular joint were calculated using ABAQUS/Standard Finite Element Software ABAQUS (ABAQUS). The composite tubes were modeled with a four-node shell finite element S4R with three integration points throughout the element thickness and the adhesive layer with an eight-node solid element C3D8R. The finite element model of the adhesive joint consists of 4824 finite elements, 5472 nodes, and 32,832 degrees of freedom (Figure 10.10).

TABLE 10.6

Optimal Values of the Design Parameters For Each Mode of the Adhesively Bonded Composite Single Lap Joint Based on Three Objective Functions (Apalak and Yildirim, 2009)

Mode	First Objective Function $c_1 = 0.5$ and $c_2 = 0.5$				Second Objective Function $c_1 = 0.0$ and $c_2 = 1.0$				Third Objective Function $c_1 = 1.0$ and $c_2 = 0.0$			
	c (mm)	α	V_f (%)	t_1 (mm)	c (mm)	α	V_f (%)	t_1 (mm)	c (mm)	α	V_f (%)	t_1 (mm)
1	50.0	6.4	70.0	0.5	10.0	86.7	40.0	0.5	50.0	0.0	70.0	2.0
2	50.0	10.2	64.2	0.5	10.0	71.1	40.0	0.5	50.0	0.0	70.0	2.0
3	50.0	17.0	70.0	1.2	10.0	58.3	40.0	0.5	50.0	8.7	70.0	2.0
4	10.0	20.7	40.0	0.6	10.0	90.0	40.0	0.5	50.0	0.0	70.0	1.5
5	50.0	19.5	70.0	0.5	13.5	67.0	40.0	0.9	50.0	0.0	70.0	2.0
6	44.5	33.9	70.0	0.7	10.0	69.7	40.0	0.5	50.0	14.4	70.0	2.0
7	50.0	3.7	55.5	2.0	10.0	69.9	40.0	0.5	50.0	5.7	70.0	2.0
8	15.8	21.5	70.0	0.6	46.4	90.0	40.0	0.5	50.0	8.7	70.0	2.0
9	10.0	6.6	40.0	0.5	22.6	39.4	40.0	0.5	50.0	6.3	70.0	2.0
10	50.0	20.9	70.0	0.5	50.0	90.0	40.0	0.5	50.0	7.8	70.0	2.0

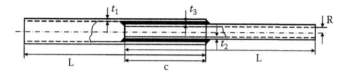

FIGURE 10.9
Adhesively bonded composite tubular single lap joint (Apalak and Yildirim, 2007).

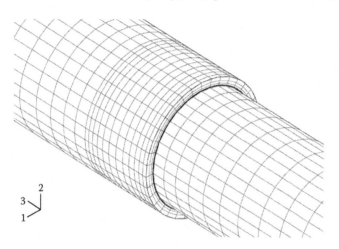

FIGURE 10.10
Mesh detail of the overlap region in an adhesively bonded composite tubular single lap joint (Apalak and Yildirim, 2007).

First, free vibration analysis was carried out for the tubular joint with tube thickness $t_1 = t_2 = 1.0$ mm, inner tube radius $R = 12.5$ mm, overlap length $c = 20$ mm, adhesive thickness $t_3 = 0.2$ mm, fiber angle $\alpha = 0°$, and fiber volume fraction $V_f = 0.635$. The first 10 natural frequencies occur in the bending, longitudinal, and right-side love mode shapes (Figure 10.11). Natural frequencies in the bending modes were measured as follows: $\omega_1 = \omega_2 = 58.8$, $\omega_3 = \omega_4 = 346.9$, $\omega_6 = \omega_7 = 953.3$, and $\omega_9 = \omega_{10} = 1727.3$ Hz. The right-side love mode shape occurs at a natural frequency $\omega_5 = 846.8$ and the longitudinal mode shape appears at a natural frequency $\omega_8 = 1400.4$ Hz. The bending mode shape is dominant and the neighboring bending modes coincide in the perpendicular radial directions. Since the mechanical and physical properties of the adhesive material, such as modulus of elasticity, Poisson's ratio, and density, have a negligible effect on both natural frequencies and modal strain energies of the composite tubular adhesive joint (Yildirim, 2006), the mechanical and physical properties of the adhesive layer and the adhesive thickness $t_3 = 0.2$ mm were kept constant throughout the study.

The effect of fiber angle α on the first 10 natural frequencies and the corresponding modal strain energies are given in Table 10.7 for fiber angle α

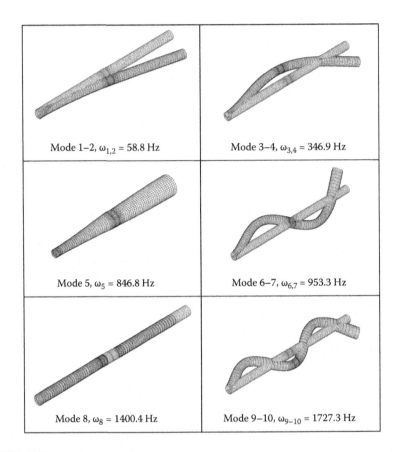

FIGURE 10.11
The first 10 natural frequencies and mode shapes of an adhesively bonded composite tubular single lap joint (t_3 = 0.2 mm, t_1 = t_2 = 1 mm, L = 250 mm, c = 20 mm, R = 12.5 mm, α = 0°, V_f = 0.635) (Apalak and Yildirim, 2007).

between 0° and 90° as the other joint dimensions are kept constant. Natural frequencies are increased by 18–78% with increasing fiber angle. In addition, modal strain energies increase considerably (by 100%). However, a good joint design can be achieved by maximizing the natural frequency, while the modal strain energy is minimized. Therefore, fiber angle α of the composite tubes appears to be a dominant design parameter. Because V_f varies from 40% to 70%, the first 10 natural frequencies increase by 25% (Table 10.8), whereas modal strain energies increase by 50–70%. V_f also appears to be another design parameter in terms of natural frequency and modal strain energy. Increasing the overlap length c by between 10 and 50 mm results in increases by 6–15% in the first natural frequencies and increases by 17–47% in the modal strain energies of the unidirectional adhesive joints with α = 0° and V_f = 0.635 (Table 10.9). The adhesive joint becomes stiffer in the overlap region with increasing overlap length, and frequencies and adhesive

TABLE 10.7

Effect of Fiber Angle (α) on (a) the First 10 Natural Frequencies and (b) the Modal Strain Energies of the Adhesively Bonded Composite Tubular Single Lap Joint (Apalak and Yildirim, 2007)

(a)

	Normalized Natural Frequencies (ω_i/ω_{max})					
Mode	$\overline{\omega}_{1,2}$	$\overline{\omega}_{3,4}$	$\overline{\omega}_5$	$\overline{\omega}_{6,7}$	$\overline{\omega}_8$	$\overline{\omega}_{9,10}$
α (°)	213.0	996.8	1046.4	2302.0	2684.4	3510.5
0	0.2821	0.3559	0.8258	0.4229	0.5322	0.5031
15	0.2766	0.3510	0.8988	0.4199	0.5245	0.5047
30	0.2721	0.3484	0.9276	0.4217	0.5130	0.5161
45	0.2903	0.3736	1.0000	0.4546	0.7156	0.5619
60	0.3611	0.4636	0.9879	0.5586	0.8948	0.6843
75	0.5725	0.7106	0.8709	0.8092	1.0000	0.9280
90	1.0000	1.0000	0.9526	1.0000	0.9473	1.0000

(b)

	Normalized Modal Strain Energies					
Mode	1.2	3.4	5	6,7	8	9,10
α (°)	29,302	763,738	1,769,804	4,823,404	8,108,014	12,569,002
0	0.0743	0.1017	0.5265	0.1559	0.3148	0.1694
15	0.1213	0.0959	0.6288	0.1289	0.3088	0.1745
30	0.0737	0.1709	0.5157	0.1892	0.2956	0.2029
45	0.1532	0.1356	0.6713	0.2463	1.0000	0.3354
60	0.2064	0.1865	1.0000	0.2895	0.4601	0.2968
75	0.3184	0.4132	0.6256	0.4986	0.9671	1.0000
90	1.0000	1.0000	0.4315	1.0000	0.8145	0.8729

TABLE 10.8

Effect of Fiber Volume Fraction (V_f) on (a) the First 10 Natural Frequencies and (b) Modal Strain Energies of the Adhesively Bonded Composite Tubular Single Lap Joint (Apalak and Yildirim, 2007)

(a)

	Normalized Natural Frequencies (ω/ω_{max})					
Mode	$\bar{\omega}_{1,2}$	$\bar{\omega}_{3,4}$	$\bar{\omega}_5$	$\bar{\omega}_{6,7}$	$\bar{\omega}_8$	$\bar{\omega}_{9,10}$
V_f (%)	**64.0**	**378.0**	**918.9**	**1036.7**	**1524.0**	**1880.4**
40	0.7515	0.7524	0.7600	0.7524	0.7516	0.7537
44	0.7721	0.7729	0.7801	0.7729	0.7721	0.7741
48	0.7951	0.7959	0.8027	0.7959	0.7951	0.7970
52	0.8212	0.8219	0.8281	0.8219	0.8213	0.8229
56	0.8510	0.8517	0.8571	0.8516	0.8511	0.8525
60	0.8854	0.8859	0.8903	0.8858	0.8854	0.8866
64	0.9254	0.9258	0.9288	0.9257	0.9254	0.9263
68	0.9728	0.9729	0.9741	0.9729	0.9728	0.9731
70	1.0000	1.0000	1.0000	1.0000	1.0000	1.0000

(b)

	Normalized Modal Strain Energies					
Mode	**1.2**	**3.4**	**5**	**6.7**	**8**	**9,10**
V_f (%)	**64.0**	**378.0**	**918.9**	**1036.7**	**1524.0**	**1880.4**
40	0.5168	0.3259	0.5262	0.3134	0.5146	0.3403
44	0.5026	0.4126	0.5616	0.4274	0.5502	0.4401
48	0.4964	0.4020	0.6022	0.5034	0.5911	0.3989
52	0.5390	0.5857	0.6492	0.3942	0.6385	0.4257
56	0.9425	0.3884	0.7040	0.4208	0.6943	0.4443
60	0.6539	0.4342	0.7690	0.5811	0.7607	0.7158
64	0.8789	0.5859	0.8473	1.0000	0.8412	0.8390
68	0.8291	0.5238	0.9433	0.5946	0.9408	0.9393
70	1.0000	1.0000	1.0000	0.6197	1.0000	0.7614

TABLE 10.9

Effect of Overlap Length (c) on (a) the First 10 Natural Frequencies and (b) the Modal Strain Energies of the Adhesively Bonded Composite Tubular Single Lap Joint (Apalak and Yildirim, 2007)

(a)

Mode	Normalized Natural Frequencies (ω_f/ω_{max})					
	$\bar\omega_{1,2}$	$\bar\omega_{3,4}$	$\bar\omega_5$	$\bar\omega_{6,7}$	$\bar\omega_8$	$\bar\omega_{9,10}$
c (mm)	67.4	393.9	903.0	1092.6	1493.1	1966.0
10	0.8556	0.8721	0.9422	0.8553	0.9422	0.8711
15	0.8729	0.8859	0.9495	0.8729	0.9494	0.8840
21	0.8909	0.9007	0.9568	0.8912	0.9568	0.8982
26	0.9095	0.9164	0.9643	0.9098	0.9643	0.9138
31	0.9285	0.9331	0.9719	0.9289	0.9719	0.9307
37	0.9482	0.9509	0.9798	0.9487	0.9798	0.9489
42	0.9684	0.9697	0.9877	0.9689	0.9877	0.9683
47	0.9893	0.9896	0.9958	0.9895	0.9958	0.9891
50	1.0000	1.0000	1.0000	1.0000	1.0000	1.0000

(b)

Mode	Normalized Modal Strain Energies					
	1,2	3,4	5	6,7	8	9,10
c (mm)	4228	123,802	1,064,074	1,208,656	2,911,140	3,899,180
10	0.5374	0.5904	0.8371	0.5777	0.8384	0.5160
15	0.5077	0.5977	0.8559	0.5532	0.8571	0.6683
21	0.5152	0.6272	0.8756	0.6222	0.8766	0.5461
26	0.5446	0.6547	0.8961	0.5819	0.8969	0.5816
31	1.0000	1.0000	0.9067	0.7561	0.9075	0.8666
37	0.8832	0.7150	0.9285	0.8403	0.9292	0.7234
42	0.9286	0.9198	0.9513	1.0000	0.9518	1.0000
47	0.7022	0.7418	0.9752	0.5924	0.9754	0.6670
50	0.8342	0.7904	1.0000	0.7438	1.0000	0.7705

deformations are increased. Therefore, the overlap length is a geometrical design parameter. In addition, increasing the inner tube thickness t_2 by between 0.53 and 2.0 mm decreases the natural frequencies by 3–42% and increases the corresponding modal strain energies (Table 10.10) by 25–85%. The inner tube becomes stiffer when the tube thickness is increased; therefore, the adhesive layer undergoes more deformation than the inner tube. The modal strain energy increases, whereas the natural frequencies of the adhesive joint decrease. Consequently, inner tube thickness is an important geometrical parameter. However, an increase of 4–47% in natural frequencies and an increase of 33–89% in modal strain energies are observed as the outer tube thickness t_1 is increased from 0.53 to 2.0 mm (Table 10.11). The adhesive layer experiences higher deformations as the composite tubes become stiffer. Increasing the inner tube radius R by between 12.5 and 30 mm results in increases of 22–57% in natural frequencies and 58–96% in modal strain energies (Table 10.12). Inner tube radius is also a geometrical design parameter of the adhesive joint.

The free vibration analysis indicates that natural frequencies and mode shapes of the composite tubular single lap joint are strongly dependent on fiber angle and volume fraction as well as geometrical parameters, such as overlap length, tube thicknesses, and inner tube radius. An optimal joint design achieving maximum natural frequency and minimum modal strain energy level can be searched within the specified limits of these design parameters, except for the mechanical properties of the adhesive layer and the adhesive thickness due to their negligible effects on natural frequencies and modal strain energies. However, the direct search for an optimal joint design requires the evaluation of 4131×10^5 models with various design variables, such as a fiber angle $0,1 \leq \alpha \leq 90°$, a fiber volume fraction $0.40,0.41 \leq V_f \leq 0.70$, a support length $10,11 \leq c \leq 50$ mm, an inner tube thickness $0.5,0.6 \leq t_2 \leq 2.0$ mm, an outer tube thickness $0.5,0.6 \leq t_1 \leq 2.0$ mm, and an inner tube radius $13,13.1 \leq R \leq 30$ mm. Each model requires a three-dimensional free vibration analysis in which the first 10 eigenvalues of each model are extracted, and the mode shape for each natural frequency and the subsequent stress and strain distributions are calculated. Therefore, the direct search for an optimal joint design is also a challenge for this problem. ANNs can be used for instantly predicting the first 10 natural frequencies and the corresponding modal strain energies for various values of the design parameters. However, this requires that an appropriate ANN model be designed and trained with a small amount of data from the free vibration analysis of the adhesive joint for random values of the design parameters. Consequently, the optimal values of effective design parameters on natural frequencies and modal strain energies can be searched using GA with minimum calculation.

For this purpose, a neural network model was established for each of the first 10 natural frequencies and each corresponding modal strain energy. To train and test these neural network models, a free vibration analysis of the adhesive joint was carried out using the finite element method for random

TABLE 10.10

Effect of Inner Tube Thickness (t_2) on (a) the First 10 Natural Frequencies and (b) the Modal Strain Energies of the Adhesively Bonded Composite Tubular Single Lap Joint (Apalak and Yildirim, 2007)

(a)

Mode	Normalized Natural Frequencies (ω_i/ω_{max})					
	$\bar{\omega}_{1,2}$	$\bar{\omega}_{3,4}$	$\bar{\omega}_5$	$\bar{\omega}_{6,7}$	$\bar{\omega}_8$	$\bar{\omega}_{9,10}$
t_2 (mm)	102.2	360.8	1005.7	1197.1	1741.9	1987.3
0.53	1.0000	0.9781	1.0000	1.0000	0.9778	1.0000
0.80	0.8611	0.9953	0.9821	0.9214	0.9948	0.9202
0.93	0.8106	0.9987	0.9769	0.8896	1.0000	0.8879
1.07	0.7682	1.0000	0.9734	0.8612	0.9804	0.8799
1.33	0.7003	0.9986	0.9673	0.8143	0.9244	0.8845
1.60	0.6479	0.9938	0.9189	0.8130	0.8777	0.8870
1.87	0.6058	0.9872	0.8777	0.8130	0.8379	0.8883
2.00	0.5876	0.9834	0.8592	0.8134	0.8201	0.8886

(b)

Mode	Normalized Modal Strain Energies					
	1,2	3,4	5	6,7	8	9,10
t_2 (mm)	2310	121,321	121,321	813,030	2,552,030	2,447,516
0.53	0.7557	0.1519	0.2192	0.6191	0.2038	0.5687
0.80	0.8687	0.3243	0.3478	0.7698	0.3143	0.7054
0.93	0.9803	0.3219	0.3424	0.8310	0.5827	0.7608
1.07	0.9516	0.6611	0.3718	0.8854	0.7766	0.4405
1.33	0.9418	0.4263	0.8532	0.5236	0.8562	0.5707
1.60	0.9319	0.5217	0.9196	1.0000	0.9214	1.0000
1.87	0.9424	0.6063	0.9753	0.7353	0.9759	0.8481
2.00	0.9429	0.6401	1.0000	0.9250	1.0000	0.8701

TABLE 10.11

Effect of Outer Tube Thickness (t_1) on (a) the First 10 Natural Frequencies and (b) the Modal Strain Energies of the Adhesively Bonded Composite Tubular Single Lap Joint (Apalak and Yildirim, 2007)

(a)

	Normalized Natural Frequencies (ω_i/ω_{max})					
Mode	$\overline{\omega}_{1,2}$	$\overline{\omega}_{3,4}$	$\overline{\omega}_5$	$\overline{\omega}_{6,7}$	$\overline{\omega}_8$	$\overline{\omega}_{9,10}$
t_1 (mm)	60.1	354.8	864.1	1022.2	1428.6	1766.0
0.53	0.5394	0.8268	0.6109	1.0000	0.7155	0.9688
0.80	0.6549	0.8974	0.7209	0.9777	0.7203	0.9844
1.07	0.7500	0.9389	0.8050	0.9656	0.8046	0.9926
1.33	0.8319	0.9654	0.8729	0.9588	0.8727	0.9969
1.60	0.9044	0.9831	0.9296	0.9549	0.9297	0.9990
1.87	0.9696	0.9954	0.9782	0.9530	0.9782	0.9999
2.00	1.0000	1.0000	1.0000	0.9525	1.0000	1.0000

(b)

		Normalized Modal Strain Energies					
Mode	t_1 (mm)	1.2	3.4	5	6,7	8	9,10
		2178	77,652	931,718	752,067	2,552,030	2,661,726
	0.53	0.2885	0.6715	0.3686	0.3824	0.1127	0.2967
	0.80	0.4257	0.8262	0.5146	0.5607	0.5138	0.4639
	1.07	0.5598	0.8714	0.6431	0.7281	0.6426	0.6301
	1.33	0.6890	0.9110	0.7578	0.8185	0.7576	0.8529
	1.60	0.8163	0.9452	0.8614	0.8213	0.8613	1.0000
	1.87	0.9382	0.9583	0.9558	0.7992	0.9558	0.7943
	2.00	1.0000	1.0000	1.0000	1.0000	1.0000	0.8001

TABLE 10.12

Effect of Inner Tube Radius (R) on (a) the First 10 Natural Frequencies and (b) the Modal Strain Energies of the Adhesively Bonded Composite Tubular Single Lap Joint (Apalak and Yildirim, 2007)

(a)

	Normalized Natural Frequencies (ω_i/ω_{max})					
Mode	$\bar{\omega}_{1,2}$	$\bar{\omega}_{3,4}$	$\bar{\omega}_5$	$\bar{\omega}_{6,7}$	$\bar{\omega}_8$	$\bar{\omega}_{9,10}$
R (mm)	138.8	735.9	864.1	1425.8	1810.4	2880.0
12.50	0.4329	0.4822	1.0000	0.6829	0.7891	0.6132
14.83	0.5105	0.5632	0.9973	0.7866	0.7885	0.6937
17.17	0.5879	0.6411	0.9954	0.8821	0.7881	0.7643
19.50	0.6645	0.7151	0.9939	0.9689	0.7877	0.8254
20.67	0.7026	0.7507	0.9933	1.0000	0.7950	0.8529
23.00	0.7782	0.8188	0.9924	0.9997	0.8541	0.9024
28.83	0.9635	0.9721	0.9906	0.9992	0.9786	1.0000
30.00	1.0000	1.0000	0.9903	0.9991	1.0000	0.9677

(b)

	Normalized Modal Strain Energies					
Mode	1.2	3.4	5	6,7	8	9,10
R (mm)	52,782	1,058,191	2,188,169	6,088,005	10,010,426	22,001,873
12.50	0.0413	0.0734	0.4258	0.1235	0.2549	0.0968
14.83	0.0682	0.1166	0.5021	0.1610	0.3019	0.1586
17.17	0.1052	0.1779	0.5788	0.2419	0.3492	0.2231
19.50	0.2519	0.2563	0.6552	0.3434	0.3962	0.3077
20.67	0.2111	0.3039	0.6936	0.6904	0.2446	0.3552
23.00	0.2692	0.4156	0.7701	0.7677	0.3253	0.4586
28.83	0.7749	0.7663	0.9616	0.9612	0.7448	0.8087
30.00	1.0000	0.9608	1.0000	1.0000	0.6881	0.5411

values of the design variables, such as a fiber angle $0 \leq \alpha \leq 90°$, a fiber volume fraction $0.40 \leq V_f \leq 0.70$, a support length $10 \leq c \leq 50$ mm, an inner tube thickness $0.5 \leq t_2 \leq 2.0$ mm, an outer tube thickness $0.5 \leq t_1 \leq 2.0$ mm, and an inner tube thickness $13 \leq R \leq 30$ mm. The input variables of the neural network models were fiber volume fraction, fiber angle, overlap length, inner and outer tube thicknesses, and outer tube radius (Figure 10.12). The training sessions were repeated for each neural network model. The proposed ANN models, in which the Levenberg-Marquardt back-propagation learning algorithm (Hagan and Menhaj, 1994) was implemented, are feedforward, two–hidden layer networks. Thirty and 10 neurons were used in the first and second hidden layers, respectively. The logarithmic sigmoid and tangent sigmoid transfer functions (Figure 10.5) are activation functions used throughout the hidden layers and the output layer. The training and testing data were normalized to a range of 0–1. The proposed neural network models were developed using the neural network toolbox of MATLAB (MATLAB). For the training and testing of proposed neural network models, the data set includes the design variables and the first 10 natural frequencies, as well as modal strain energies calculated based on the free vibration analysis of the adhesive joint for 5000 random values of the design variables. The neural network model was trained with (an input data) 4500 patterns of the data set, and its prediction performance was tested with (testing data) 500 patterns randomly selected from the data set. The performance of the training session was evaluated with the mean square error reduced to an order of 10^{-8}.

The adhesively bonded composite tubular single lap joint experiences mostly bending modes, and fiber angle α, overlap length c, inner tube radius R, and outer tube thickness t_1 affect both the natural frequencies and the corresponding modal strain energies. The proposed neural network models were

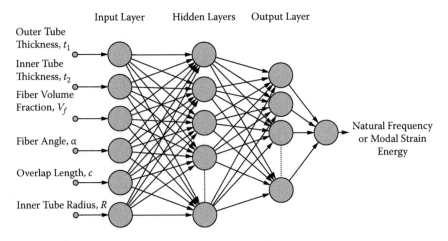

FIGURE 10.12
ANN model for the adhesively bonded composite tubular lap joint (Apalak and Yildirim, 2007).

used to investigate the couple effect of these dominant design parameters on natural frequencies and modal strain energies for a fiber volume fraction V_f = 0.635. As fiber angle α is increased, all natural frequencies exhibit considerable increases (by 150–250%); in particular, this effect becomes more evident when α = 40° (Figure 10.13). Fiber angle α appears as a more dominant design parameter in comparison with overlap length c, resulting in small increases in the first 10 natural frequencies. Inner tube radius R causes marked increases in the first 10 natural frequencies (Figure 10.14) in comparison with overlap length c. However, the effect of inner tube radius R disappears when R = 20 mm for modes 6 and 9. In addition, outer tube thickness t_1 is less effective on first natural frequencies than inner tube radius R (Figure 10.15).

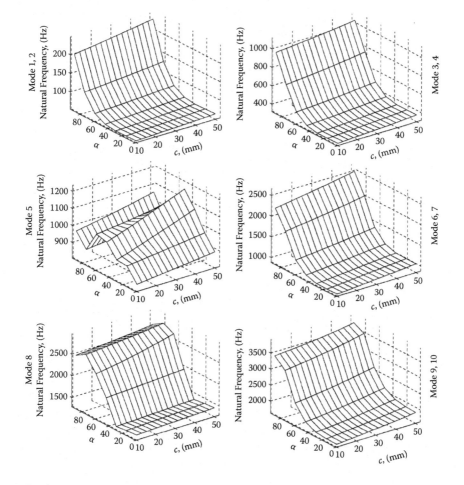

FIGURE 10.13
Combined effect of overlap length (c) and fiber angle (α) on the first 10 natural frequencies of the adhesively bonded composite tubular single lap joint ($t_1 = t_2 = 1$ mm, $R = 12.5$ mm, $V_f = 0.635$) (Apalak and Yildirim, 2007).

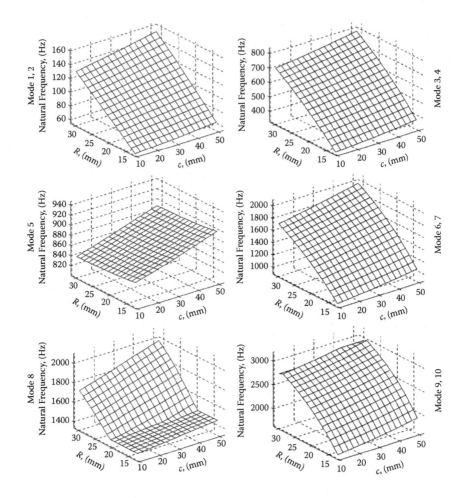

FIGURE 10.14
Combined effect of overlap length (c) and inner tube radius (R) on the first 10 natural frequencies of the adhesively bonded composite tubular single lap joint ($t_1 = t_2 = 1$ mm, $\alpha = 0°$, $V_f = 0.635$) (Apalak and Yildirim, 2007).

It is apparent that fiber angle α and inner tube radius R are the more dominant design parameters on the natural frequencies of the adhesively bonded composite tubular single lap joint. Thus, the first 10 natural frequencies become maximal when fiber angle α, inner tube radius R, overlap length c, and outer tube thickness t_1 are maximal. An optimal design for the adhesively bonded composite tubular lap joint requires that natural frequencies be maximized for each mode and corresponding modal strain energies be minimized. GA was used to search for a most appropriate set of design parameters of the adhesive tubular joint based on the evaluation of an objective function including the natural frequency and modal strain energy values. Neural network models were used to instantly predict the

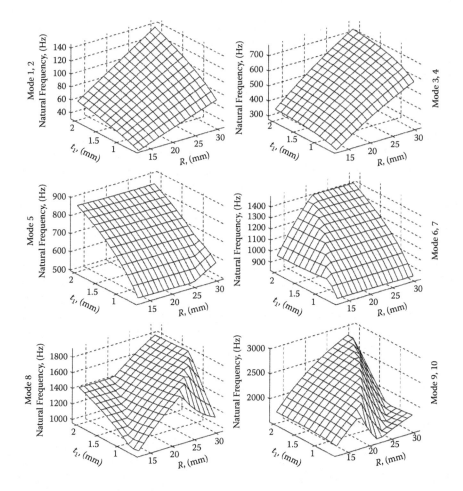

FIGURE 10.15
Combined effect of inner tube radius (R) and outer tube thickness (t_1) on the first 10 natural frequencies of the adhesively bonded composite tubular single lap joint ($t_2 = 1$ mm, $c = 20$ mm, $\alpha = 0$, $V_f = 0.635$) (Apalak and Yildirim, 2007).

first 10 natural frequencies and the corresponding modal strain energies for a new set of design parameters instead of the finite element method, which requires a time-consuming calculation phase. GA and ANN toolboxes of MATLAB (MATLAB) were used to solve the optimal joint problem defined as follows.

Find the fiber volume fraction V_f, fiber angle α, overlap length c, inner tube thickness t_2, outer tube thickness t_1, and inner tube radius R that will minimize the objective function

$$f(\omega_i, U_i) = -c_1 \times c_2 \times U_i \tag{10.26}$$

where $c_1 = c_2 = 0.0{-}1.0$, ω_i is the *i*th natural frequency, and U_i is the corresponding strain energy, subject to the following constraints: $0.40 \leq V_f \leq 0.70$, $0 \leq \alpha \leq 90°$, $10 \leq c \leq 50$ mm, $0.5 \leq t_2 \leq 2.0$ mm, $0.5 \leq t_1 \leq 2.0$ mm, and $13 \leq R \leq 30$ mm.

Three objective functions were defined as follows: (1) natural frequency and modal strain energy were equally weighted ($c_1 = c_2 = 0.5$), (2) only the modal strain energy was minimized ($c_1 = 0.0$ and $c_2 = 1.0$), and (3) only the natural frequency was maximized ($c_1 = 1.0$ and $c_2 = 0.0$); the optimal values of the joint design parameters are given in Table 10.13. When only the natural frequencies are maximized, the overlap length c, inner and outer tube thicknesses t_2 and t_1, and inner tube radius R are maximized as much as possible. Fiber volume fraction V_f reaches 70% and fiber angle α is maximal for the vlow frequencies. When modal strain energy is minimized, overlap length and fiber angle are maximized, whereas inner tube radius and tube thicknesses are minimized. The volume fraction gets a value between 40% and 70%. The design parameters behave differently depending on both objective functions. Thus, the design parameters reach completely opposite values for either minimizing the modal strain energies or maximizing the natural frequencies. The first objective function equally balances our expectations from the adhesive joint behavior. All design parameters are maximized as much as possible except for inner tube thickness, and the fiber volume fraction becomes optimal for 70%.

10.6 Conclusions

This study investigates the design parameters affecting the free vibration behaviors of adhesively bonded clamped-free composite single lap and tubular lap joints formed bonding unidirectional plates and tubes via an adhesive layer. The joint design parameters were determined such that maximum natural frequency and minimum modal strain energy conditions can be satisfied using GA combined with ANN. In the case of adhesively bonded clamped-free composite single lap joints,

- The mechanical properties of the adhesive material and the adhesive thickness had a minor effect on the natural frequencies and the modal strain energies of both joints.
- The adhesively bonded single lap joint undergoes transverse bending and torsional modes for the first six natural frequencies.
- As the fiber volume fraction, overlap length, and plate thickness are increased, the frequencies and modal strain energies increase markedly, but decrease with increasing fiber angle.

TABLE 10.13

Optimal Values of the Six Design Parameters (c, R, V_f, α, t_2, t_1) of the Adhesively Bonded Composite Tubular Single Lap Joint (Apalak and Yildirim, 2007)

Mode	First Objective Function $c_1 = 0.5$ and $c_2 = 0.5$						Second Objective Function $c_1 = 0.0$ and $c_2 = 1.0$						Third Objective Function $c_1 = 1.0$ and $c_2 = 0.0$					
	c (mm)	R (mm)	V_f (%)	α (°)	t_1 (mm)	t_2 (mm)	c (mm)	R (mm)	V_f (%)	α (°)	t_1 (mm)	t_2 (mm)	c (mm)	R (mm)	V_f (%)	α (°)	t_1 (mm)	t_2 (mm)
1	45	23	40	90	0.5	2.0	22	13	42	23	0.5	0.5	50	30	70	90	1.1	2.0
2	17	25	63	90	0.5	2.0	26	13	40	6	0.5	0.5	50	21	70	90	0.5	2.0
3	50	25	70	55	1.0	2.0	14	15	55	87	0.5	1.0	50	27	70	62	2.0	2.0
4	50	30	70	45	1.3	2.0	46	16	40	90	0.5	1.6	50	30	70	53	2.0	2.0
5	47	21	70	90	1.7	2.0	11	21	51	45	0.5	2.0	50	19	70	34	0.9	2.0
6	50	21	70	51	1.6	2.0	24	13	60	45	1.3	2.0	50	22	70	51	2.0	2.0
7	50	16	70	90	2.0	1.9	21	13	43	90	0.9	2.0	50	23	70	42	2.0	2.0
8	50	15	70	51	1.2	0.7	10	13	59	46	0.5	1.8	50	13	70	79	2.0	1.6
9	50	13	70	70	1.4	2.0	21	15	70	90	0.8	1.3	50	22	70	23	2.0	2.0
10	50	19	70	0	0.6	0.9	31	30	70	90	1.3	1.3	50	26	70	11	2.0	2.0

- Depending on the minimization of the modal strain energy or the maximization of the natural frequencies, the design parameters yield different optimal values since natural frequency and modal strain energy have opposite responses to the design parameters. When only natural frequencies are maximized, the overlap length c, plate thickness t_1, and fiber volume fraction V_f become maximal, and fiber angle α approaches zero or becomes smaller than 14°. When only modal strain energies are minimized, the overlap length, plate thickness, and fiber volume fraction are minimized, whereas fiber angle is maximized.

- When natural frequency and strain energy are weighted equally, the overlap length and fiber volume fraction become maximal, whereas fiber angle and plate thickness are minimal.

For the adhesively bonded clamped-free composite tubular lap joint,

- The bending modes become dominant for the first 10 natural frequencies.

- Increasing the fiber volume fraction, fiber angle, inner tube radius, overlap length, and outer tube thickness lead to increases in natural frequencies and modal strain energies, whereas increasing the inner tube thickness reduces natural frequencies but increases modal strain energies.

- On natural frequencies and modal strain energies indicating a larger fiber angle than 20°, the inner tube radius is a more effective design parameter than the overlap length and the outer tube thickness.

- Similarly, the maximization of only the natural frequencies requires the overlap length c, inner and outer tube thicknesses t_2 and t_1, inner tube radius R, and fiber angle α to be maximized, and a fiber volume fraction V_f of 70%. The minimization of the modal strain energy means that the overlap length and fiber angle are maximal, whereas the inner tube radius and the tube thicknesses are minimal. The volume fraction stays between 40% and 70%. When natural frequencies and modal strain energies are equally weighted, all design parameters reach their upper bounds except for inner tube thickness and fiber volume fraction (70%).

References

ABAQUS/Standard (Version 6.5), Finite element software, www.abaqus.com.

Adams, R. D., and Wake, W. C. 1984. *Structural adhesive joints in engineering.* London: Elsevier Applied Science.

Apalak, M. K., and Yildirim, M. 2007. Free vibration analysis and optimal design of a cantilevered adhesively bonded composite tubular single lap joint. *Polymers and Polymer Composites* 15:489–506.

Apalak, M. K., and Yildirim, M. 2009. Free vibration analysis and optimal design of a clamped-free single lap joint with unidirectional laminated narrow plates. *Journal of Thermoplastic Composites* 22:183–211.

Cook, R. D. 1981. *Concepts and applications of finite element analysis,* 2nd ed. New York: John Wiley and Sons.

Doebling, S. W., Hemez, F. M., Peterson, L. D., and Farhat, C. 1997. Improved damage location accuracy using strain energy based on mode selection criteria. *AIAA Journal* 35:693–699.

Goldberg, D. E. 1989. Genetic algorithms in search, optimization and machine learning. Boston: Kluwer Academic Publishers.

Goldberg, D. E. 2002. *The design of innovation: lessons from and for competent genetic algorithms.* Reading: Addison-Wesley.

Grimes, R. G., Lewis, J. G., and Simon, H. D. 1994. A shifted block lanczos algorithm for solving sparse symmetric generalized eigenproblems. *SIAM Journal Matrix Analysis Applications* 15:228–272.

Hagan, M. T., and Menhaj, M. 1994. Training feed-forward networks with the Marquardt algorithm. *IEEE Transactitons on Neural Networks* 5:989–993.

He, X., and Oyadiji, S. O. 2001. Influence of adhesive characteristics on the transverse free vibration of single lap-jointed cantilevered beams. *Journal of Materials Processing Technology* 119:366–373.

Holland, J. H. 1975. *Adaptation in natural and artificial systems.* Ann Arbor: University of Michigan Press.

Jaishi, B., and Ren, W. X. 2005. Structural finite element model updating using ambient vibration test results. *Journal of Structural Engineering ASCE* 131:617–628.

Jaishi, B., and Ren, W. X. 2007. Finite element model updating based on eigenvalue and strain energy residuals using multiobjective optimisation technique. *Mechanical Systems and Signal Processing* 21:2295–2317.

Kaya, A., Tekelioglu, M. S., and Findik, F. 2004. Effects of various parameters on dynamic characteristics in adhesively bonded joints. *Materials Letters* 58:3451–3456.

Ko, T. C., Lin, C. C., and Chu, R. C. 1995. Vibration of bonded laminated lap-joint plates using adhesive interface elements. *Journal Sound and Vibration* 184:567–583.

Lim, T. W., and Kashangaki, T. A. L. 1994. Structural damage detection of space truss structures using best achievable eigenvectors. *AIAA Journal* 32:1049–1057.

Lin, C. C., and Ko, T. C. 1997. Free vibration of bonded plates. *Computers and Structures* 64:441–452.

Loy, C. T., Lam, K. Y., and Reddy, J. N. 1999. Vibration of functionally graded cylindrical shells. *International Journal of Mechanical Sciences* 41:309–324.

MATLAB (ver 7.0), The MathWorks Inc., Natick, MA.

Michalewicz, Z. 1999. *Genetic algorithms + data structures = evolution programs.* Berlin: Springer-Verlag.

Pradhana, S. C., Loy, C. T., Lam, K.Y., and Reddy J. N. 2000. Vibration characteristics of functionally graded cylindrical shells under various boundary equations. *Applied Acoustics* 61:111–129.

Rao, S. S. 1989. *The finite element method in engineering.* Oxford: Pergamon Press.

Reddy, J. N. 1997. *Mechanics of laminated composite plates: theory and analysis.* Boca Raton: CRC Press.

Saito, H., and Tani, H. 1984. Vibration of bonded beams with a single lap adhesive joint. *Journal of Sound and Vibration* 92:299–309.

Shi, Z. Y., and Law, S. S. 1998. Structural damage localization from modal strain energy change. *Journal of Sound and Vibration* 218:825–844.

Shi, Z. Y., Law, S. S., and Zhang, L. M. 2000. Structural damage detection from modal strain energy change. *Journal of Engineering Mechanical ASCE* 126:1216–1223.

Simpson, P. K. 1990. *Artificial neural systems: foundations, paradigms, applications and implementations*. New York: Pergamon Press.

Stubbs, N., and Kim, J. T. 1996. Damage localization in structures without baseline modal parameters. *AIAA Journal* 34:1644–1649.

Taleghani, B. K., and Pappa, R. S. 1996. U. S. Army Research Laboratory Technical Report 1288, Langley Research Center, Hampton, VA.

Tong, L., and Steven, G. P. 1999. *Analysis and design of structural bonded joints*. New York: Kluwer Academic Publishers.

Vaziri, A., and Nayeb-Hashemi, H. 2002. Dynamic response of tubular joints with an annular void subjected to a harmonic axial load. *International Journal Adhesion and Adhesives* 22:367–373.

Yang, S., Gibson, R. F., Gu, L., and Chen, W. H. 1998. Modal parameter evaluation of degraded adhesively bonded composite beams. *Composite Structures* 43:79–91.

Yildirim, M. 2006. Graduate school of natural and applied sciences. M.S. thesis, Erciyes University, Turkey.

Yuceoglu, U., and Ozerciyes, V. 2000. Sudden drop phenomena in natural frequencies of composite plates or panels with a central stiffening plate strip. *Computers and Structures* 76:247–262.

Yuceoglu, U., and Ozerciyes, V. 2003. Free flexural vibrations of orthotropic composite base plates or panels with a bonded noncentral (or eccentric) stiffening plate strip. *ASME Journal of Vibration and Acoustics* 125:228–243.

Yuceoglu, U., and Ozerciyes, V. 2005. Free vibrations of bonded single lap joints in composite shallow cylindrical shell panels. *AIAA Journal* 43:2537–2548.

Yuceoglu, U., Toghi, F., and Tekinalp, O. 1996. Free bending vibrations of adhesively bonded orthotropic plates with single lap joint. *ASME Journal of Vibration and Acoustics* 118:122–134.

11

Determining Initial Design Parameters by Using Genetically Optimized Neural Network Systems

I. N. Tansel, M. Demetgul, and R. L. Sierakowski

CONTENTS

ABSTRACT   The genetically optimized neural network system (GONNS) has been introduced for automated modeling and optimization of systems when only experimental data or observations are available. GONNS represents the system with multiple artificial neural networks after a training process. The optimum solution is calculated by using one or more genetic algorithms operating simultaneously. The GONNS does not require the user to provide any analytical or empirical models. GONNS was implemented for optimization of three systems in this paper. First, the estimation accuracy of the GONNS was evaluated by using the data of an analytical expression with three inputs. Second, the feasibility of the GONNS was studied for control of the impact resistance of composite materials by applying electricity. Third, selection of optimal operating conditions and material by using the Composite Material Selection Advisor package was demonstrated. Composite Material Selection Advisor uses the GONNS for modeling and optimization. It has additional programs for estimation of the complexity index of parts from the STL files of the commercial computer-aided design/ computer-aided manufacturing programs. If necessary, the complexity is corrected by removing the vertical holes that will be drilled after the manufacturing operation. All three implementations demonstrated the feasibility of the GONNS.

11.1   Introduction

Utilization of composite materials for the demanding civilian and military applications has been expanding rather quickly in the past few decades. This expansion is a direct result of their high strength-to-weight ratio, corrosion resistance, and controllability of characteristics in different directions. In this, relatively new, quickly developing, and very challenging field, engineers need more efficient computational tools to determine the design parameters. In this paper, research on a humanlike decision-making tool is outlined and its implementation to composite material manufacturing is discussed. Artificial neural networks (ANNs) were developed to simulate the estimation capability of biological systems. Once a generic estimation program is prepared to represent a system, many optimization techniques may be used to find the best possible solution. Genetic algorithm (GA) is one of the best techniques for optimization of a system represented by ANNs. The genetically optimized neural network system (GONNS) was developed by using multiple ANNs and GAs. The GONNS may be used for representation and optimization of many characteristics of composite materials based on experimental data.

In this paper, three implementations of the GONNS will be presented. First, the accuracy of the GONNS will be studied by using the data generated by an analytical equation. Second, implementation of GONNS will be proposed for the determination of plate thickness and applied current combination for the control of the impact resistance characteristics of hybrid materials. Third, a comprehensive package, Composite Material Selection Advisor (CoMaSA) will be discussed for the selection of composite material and optimal processing parameters. CoMaSA uses GONNS after it automatically estimates the complexity of a part from the STL files of the computer-aided design (CAD) programs.

One of the best-known ANN algorithms is the back-propagation method (Rumelhart et al., 1986; Dayhoff, 1990; Hecht-Nielsen, 1990; DARPA, 1988; Chryssolouris et al., 1992). Back-propagation algorithms represented many systems with good accuracy even with limited data (Tansel, 1990; Tansel et al., 1997). The source code of the algorithm may be found in many books and web sites (Masters, 1995). Linear and nonlinear systems may be numerically modeled using ANNs. Once the model has been established, optimal solutions can be found by various optimization methods. GA is relatively slow but works well with the ANNs without converging to local minimums as easily as many other methods (Goldberg, 1989; Carroll, 1996; Winter et al., 1996; Ladd, 1996; Ko and Kim, 1998; Tansel et al., 1999). GONNS was developed by using ANNs and GAs together. This approach was introduced, for the first time, by Tansel et al. (1999) at the Artificial Neural Networks in Engineering Conference (ANNIE '99). His group, as well as many other researchers, has used this combination for many applications (Yang et al., 2003a, 2003b; Tansel et al., 2006).

The effect of the electric current on the low-velocity impact response of carbon fiber polymer matrix was studied previously (Sierakowski et al., 2008).

Researchers applied current at three different levels to unidirectional and cross-ply composite plates for short time frames and evaluated the impact resistance of the material. The study indicated that the impact resistance of the composite material improved with the magnitude of the current as long as the electricity is applied for a short time. When the current was applied longer, there was no significant benefit to the impact resistance. Some of the data presented by Sierakowski et al. (2008) are used in this paper. To demonstrate the implementation of the GONNS, additional data were generated by assuming that the maximum load and absorbed energy would be proportional to the thickness of the material. The relationship among the impact resistance, applied current, and part thickness depends on the components of the composite material and manufacturing techniques. We believe more complex relationships among the three parameters would be observed for most composites. GONNS would learn the characteristics of such data and perform the optimization without any problems. It is recommended that the readers collect their own experimental data if they consider similar hybrid material applications.

Many cost estimation methods have already been developed for composite material manufacturing (Kaplan and Atkinson, 1989; Butler, 1994;

Krebs et al., 1997; Eaglesham, 1998; Venkataraman, 2000; Rohani and Dean, 1996; Hess and Romanoff, 1987; LeBlanc, 1976; Ramkumar et al., 1991; Mabson et al., 1994; Wong et al., 1992; Gutowski et al., 1994; Niazi et al., 2006; Tse, 1990; Muter, 1993; Kim, 1993; Pugh, 1991; Pearce, 1989; Boothroyd and Dewhurst, 1991; Li et al., 1997; Girivasan et al., 2000). Most of them used their own databases to correlate the design parameters of the parts and the manufacturing cost. The CoMaSA (Yang et al., 2003b) package was prepared to select the best material and process parameters similar to what an experienced engineer would do. The package calculates the complexity index from the CAD files. If necessary, CoMaSA corrects the index by identifying the round holes that may be drilled after the composite part is manufactured. The optimization is performed using GONNS (Yang et al., 2003a).

11.2　Theoretical Background

In this section, the ANN and GA are briefly presented. GONNS, estimation of the cost of composite material manufacturing, and previous studies on the automation of the complexity estimation are discussed.

11.2.1　ANNs and GAs

Back-propagation type ANN (Rumelhart et al., 1986) may be used for classification or mapping. The network calculates the output by using neurons located at the input, output, and hidden layer(s). One hidden layer is sufficient for most applications. The numbers of the neurons at the input and the output layers depend on the problem. The user determines the number of neurons at the hidden layer(s) based on experience and/or by comparing the performance of several models with different number of nodes. Neurons of each layer are connected to the neurons of the following layers (Figure 11.1). The neural network learns the relationship between the input(s) and the output(s) by repeating the forward and backward propagation steps. Each neuron of a layer is generally connected to the neurons in the proceeding layer. The theory and the algorithm of the back-propagation method were presented by Rumelhart et al. (1986).

　GA was developed based on the biological evolution principles including natural selection, and survival of the fittest methods (Goldberg, 1989; Carroll, 1996; Winter et al., 1996; Ladd, 1996) (Figure 11.2). The GA program works with a binary string called chromosome and tries to perfect it to make the objective function minimum or maximum. The chromosome represents all the considered parameters and switches according to the desired resolution. After the user selects the population size, the number

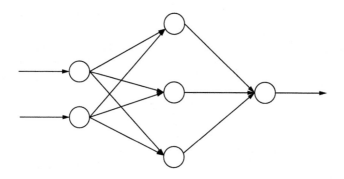

FIGURE 11.1
Diagram of back-propagation type neural network with two inputs, three hidden nodes, and one output.

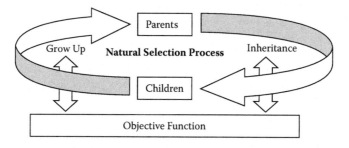

FIGURE 11.2
Simplified diagram of the operation of the genetic algorithm (Goldberg, 1989; Carroll, 1996; Winter et al., 1996; Ladd, 1996).

of children for each set of parents, and the probability of mutation, the GA modifies the digits of the binary string by mimicking the biological evaluation. The steps for mimicking the biological evaluation are as follows: (1) selection of the mating parents; (2) selection of the hereditary chromosome from parents; (3) gene crossover, gene mutation; (4) creation of the next generation.

11.2.2 Genetically Optimized Neural Network System

The GONNS (Yang et al., 2003a) uses one or multiple ANNs to represent the considered system. In our implementation, the ANNs have only one output for best possible accuracy. Depending on the system, a GA may be used with one or multiple ANNs (Figure 11.3). For complex problems, multiple clusters are optimized at the same time (Figure 11.4). The clusters have their own ANNs and GAs. After the training of the ANNs is completed, the GAs search for the best solution by using the outputs of the ANNs at different operating conditions.

In our implementation, instead of the conventional objective function with weights for the outputs of the model(s), only one parameter is minimized or maximized. The GONNS kept the other parameters at the desired ranges determined by the operator. The program uses heavy penalties to keep the variables within the determined boundaries.

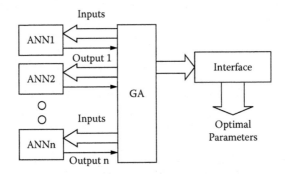

FIGURE 11.3
Single cluster GONNS with *n* ANNs and single GA.

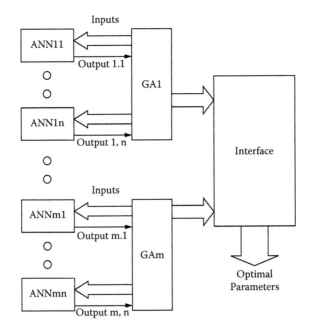

FIGURE 11.4
GONNS with *m* clusters. Each cluster has *n* ANNs and one GA.

11.2.3 Cost Estimation in Composite Material Manufacturing

Composite materials have been widely used for demanding military and civilian equipment manufacturing. Relative to conventional manufacturing with metals and polymers, production of machine parts using composite materials is much more complex. Depending on the processing parameters, selected resin, reinforcement material, and its orientation, very different characteristics may be obtained. Estimation of the material performance, manufacturing time, and cost are still challenging problems of composite material manufacturing.

Manufacturing cost (Kaplan and Atkinson, 1989; Butler, 1994; Krebs et al., 1997; Eaglesham, 1998) may be calculated by using either the volume-based methods or activity-based costing. Generally for the composite parts, activity-based costing is more accurate since manufacturing is labor intensive. Another option is using the parametric cost models (PCMs) or manufacturing process cost models (MPCMs) (Venkataraman, 2000; Rohani and Dean, 1996). One of the PCM methods is the Development and Procurement Costs for Aircraft IV (Hess and Romanoff, 1987). Some of the MPCM methods are Advanced Composite Cost Estimating Model (LeBlanc, 1976), Manufacturing Cost Model for Composites (Ramkumar et al., 1991), and Composite Optimization Software for Transport Aircraft Design Evaluation (Mabson et al., 1994). Most of the MPCM methods are developed using the databases of specific companies and their performance is limited when they are used by other groups without extensive modifications (Eaglesham, 1998; Venkataraman, 2000).

Some of the other approaches include the Totally Integrated Manufacturing Cost Estimating System (TIMCES), which separately estimates the cost of labor (Wong et al., 1992), machinery, operations, material, and overheads. To improve the accuracy of the models, the complexity of the part (Gutowski et al., 1994) was included in the model. A decision support system was proposed (Eaglesham, 1998) by using the intelligent search methods. Many cost estimation methods are discussed by Niazi et al. (2006).

11.2.4 Complexity Estimation

The complexity of a part directly affects the manufacturing cost and production time. For accurate cost and manufacturing time estimation, the complexity of the part should be calculated with minimal operator involvement. Advanced Composite Cost Estimating Model (LeBlanc, 1976) used correction factors that were obtained by considering the part geometry. Bending of fibers in composite materials (Tse, 1990), a number of nonredundant dimensions on a part drawing (Muter, 1993), and shear and ply stacking (Kim, 1993) were also used for the estimation of complexity. The level of discontinuity (Pugh, 1991), the number of dimensions in the drawing (Pearce, 1989), and symmetry (Boothroyd and Dewhurst, 1991) were used to estimate the part complexity. Consideration of the part geometry improved the accuracy of the cost estimation for machine assembly and mold-making process (Pugh, 1991; Pearce, 1989; Boothroyd and Dewhurst, 1991). Linking complexity to known cost of similar equipment (Niazi et al., 2006) and consideration of the angle(s) of the curvatures (Li et al., 1997) are some of the relatively recent approaches for correlation of the part geometry and cost.

11.3   Implementations of the GONNS

In this paper, GONNS was used for modeling and optimization of three systems. The first two were the straightforward implementation of the GONNS. The data were generated by an analytical model and used for training of the ANNs in the first implementation. The estimations of the GONNS were compared with theoretical maximum and coordinates. Semiexperimental impact-resistant control data of composite plates were used in the second implementation. The specialized cost estimation package CoMaSA was used in the third implementation. The cost estimation package had additional tools for automatic estimation of the part complexity from the STL files of the commercial CAD packages.

11.3.1   Evaluation of the Accuracy of the GONNS by Data of an Analytical Equation

A three-input-and-one-output system was considered. The system had the following relationship between the three inputs and the output:

$$f(x,y,z) = \sin(z)e^{-(x^*x+y^*y)} \tag{11.1}$$

For training, 3200 cases were prepared by using Equation (11.1). Backpropagation type ANN with 12 nodes at a single hidden layer was trained. The maximum function was searched by GONNS, and it used ANNs for the calculation of the values.

11.3.2   Determination of the Optimal Current and Thickness of the Carbon Fiber Polymer Matrix for Desired Impact Resistance Characteristics Using GONNS

Sierakowski et al. (2008) used 4.5 mm (0.18 in.) thick, 152.4 by 152.4 mm (6 × 6 in.) 32-ply carbon fiber polymer matrix composite plates to investigate the effect of short- and long-term electric currents on their impact resistances. They collected their test data at three different current levels. In this study, we wanted to optimize at least two parameters with GONNS. Since no experimental data were available, we assumed a linear relationship between the thickness and the outputs (maximum load and absorbed energy). The relationship between the plate thickness and the impact resistance depends on the involved materials and the manufacturing process. More complex relationships are expected for most composites; however, the neural networks of the GONNS would also learn such a relationship once experimental data become available.

The diagram of the considered system is presented in Figure 11.5. One of the ANNs was trained to learn the relationship between the two inputs (thickness and applied current) and maximum load. The other ANN learned the relationship between the same inputs and the absorbed energy. For the training of neural networks, the data were taken selectively from Sierakowski et al. (2008) for demonstration purposes, and additional points were interpolated based on the assumption of a linear relationship between the thickness and the ANN outputs. The relationship between the inputs and the outputs are presented in Figures 11.6 and 11.7.

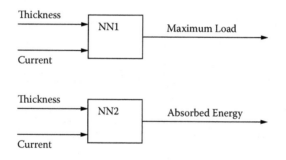

FIGURE 11.5
GONNS for the optimization of applied current–plate thickness combination.

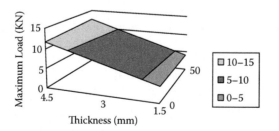

FIGURE 11.6
Simulated data for the maximum load.

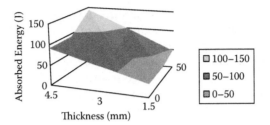

FIGURE 11.7
Training data for the simulated absorbed energy.

11.3.3 Selection of the Optimal Material and Operating Conditions

To help the operator select the best material and operating conditions, CoMaSA was developed (Yang et al., 2003b). This package has the following three programs written in Visual Basic:

- Complexity estimation from the STL files of the commercial CAD packages
- Complexity estimation correction for holes
- GONNS for optimization

11.4 Composite Material Selection Advisor

The CoMaSA (Yang et al., 2003b) was prepared to advise the engineer about the best composite material and operating conditions with minimum operator input (Figure 11.8). First, the complexity of the part is automatically calculated from the STL files of the commercial CAD packages. There are two programs for this task. The first one calculates the complexity by using all the triangles listed in the STL file. The second one identifies the vertical holes that will be drilled after the manufacturing of the composite part. The complexity index is calculated without considering the triangles on the surfaces of the vertical holes. The size and location of the holes are reported to the user to let him/her calculate the cost of the matching operation separately. In the following sections, calculation of the complexity index, removal of the vertical holes, and the considered simulation case will be presented.

11.4.1 Automated Complexity Estimation

STL format is one of the most widely used, part geometry exchange, formats between the commercial CAD packages and utility programs of the rapid prototyping machines. The CAD packages cover the surface of the part

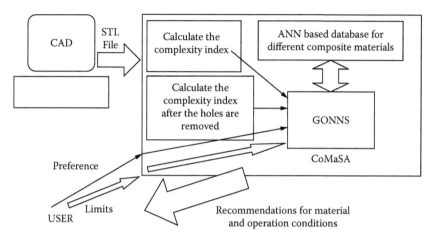

FIGURE 11.8
Diagram of the complete CoMaSA package.

with triangles according to the desired resolution and list the information in STL files. For each triangle, the coordinates of the vertices and a vector normal to the surface are listed. In this study, the part programs and their STL files were prepared using the Parametric Corporation's Pro-Engineer (Pro-E) package. The complexity estimation program calculates the angles between the each neighboring triangles and converts it into an index value using the following equation (Yang et al., 2003b; Girivasan et al., 2000):

$$\Phi = \Sigma \theta_i L_i (\text{Part}) / \Sigma \theta_k L_k \text{ (Cube having the same surface area as the part)} \quad (11.2)$$

where the angle between the normal vectors of two adjacent triangles is represented by θ_i, the length of the common side of the triangles is L_i, and the calculated $\Sigma \theta_i L_i$ for the part is divided by the $\Sigma \theta_k L_k$. $\Sigma \theta_k L_k$ is the similar value for the triangles covering the face of a cube. The complexity index of a cube is assumed to be "1" and used as the reference. The complexity of the part will be the lowest one if the normal of the triangles are parallel to each other. The complexity will increase for the larger angles. The algorithm reduces the complexity for the 90° angles since the manufacturing of the parts with perpendicular corners are simplest. A simplified algorithm of the program is presented in Figure 11.9.

11.4.2 Automated Complexity Estimation after Hole Removal

The STL file has a list of the coordinates of the vertices of all triangles, which cover the surface of the part and their normal vectors. If a part has any holes, the triangles, which cover the whole surface, are also added to the list of the STL file and the complexity index drastically increases.

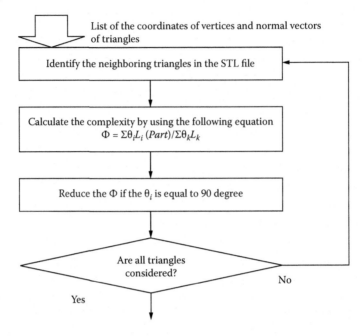

FIGURE 11.9
Simplified diagram of the complexity estimation algorithm.

Generally, holes are drilled after the composite materials are manufactured. If the hole(s) are to be drilled later, they have to be identified and costs of the composite part manufacturing and machining operations should be separately calculated.

To identify the triangles covering the vertical holes, a horizontal test plane is considered (Yang et al., 2003b; Singh, 2002) and the part is assumed to be cut with this test plane (Figure 11.10). The intersections of the sides of all the triangles covering the part surface and the test plane are calculated.

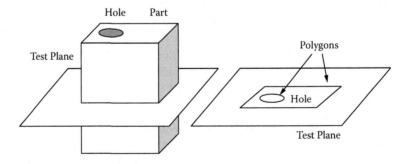

FIGURE 11.10
Intersection of the part with a horizontal test surface to identify the vertical holes (left). The test surface with polygons corresponding to the surfaces of the part and the hole (right).

Two intersection points are found between the horizontal plane and the triangles if they intersect. The triangles covering the surfaces of the part may or may not intersect the test plane. Once all the intersection points are calculated, there is at least one enclosed polygon that corresponds to the intersection of the outer surface of the part with the test plane. Additional enclosed polygons indicate the intersections of the surfaces of the holes or openings with the test plane. After the polygons of the holes are identified, the program tests them for roundness. If round holes are found, their centers and diameters are estimated. The algorithm of the program is presented in Figure 11.11.

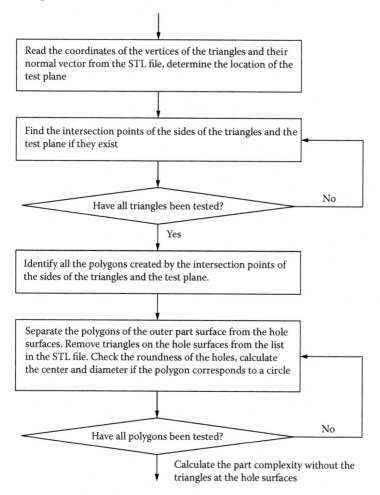

FIGURE 11.11
Simplified algorithm for the removal of triangles on hole surfaces from the original list of the STL file.

The program calculates the complexity index of the part by considering the angles between the adjacent triangles as outlined in the previous section after the triangles at the surfaces of the holes are removed from the original STL file.

11.4.3   Considered Simulation Case

In our study, the manufacturing time, cost, and strength of three composite materials were considered (Yang et al., 2003b). The following nine equations were assumed to represent these relationships (Yang et al., 2003b):

Material 1:
$$\text{Time} = 0.113 f^{0.85} v^{0.65} c^{1.15} \text{ (h)} \tag{11.3}$$
$$\text{Cost} = 3.08 f^{0.8} v^{0.7} c^{1.1} \text{ (\$)} \tag{11.4}$$
$$\text{Strength} = 226 f^{\,0.8} \text{ (kPa)} \tag{11.5}$$

Material 2:
$$\text{Time} = 0.145 f^{0.82} v^{0.67} c^{1.27} \text{ (h)} \tag{11.6}$$
$$\text{Cost} = 3.15 f^{0.73} v^{0.68} c^{1.21} \text{ (\$)} \tag{11.7}$$
$$\text{Strength} = 332 f^{0.82} \text{ (kPa)} \tag{11.8}$$

Material 3:
$$\text{Time} = 0.152 f^{0.84} v^{0.69} c^{1.32} \text{ (h)} \tag{11.9}$$
$$\text{Cost} = 3.25 f^{0.82} v^{0.66} c^{1.31} \text{ (\$)} \tag{11.10}$$
$$\text{Strength} = 345 f^{0.85} \text{ (kPa)} \tag{11.11}$$

where v is volume of the part, c is the complexity, and f denotes the fiber volumes. The actual characteristics of the composite materials could not be used since material suppliers and manufacturers did not want to share their confidential data.

11.5   Results and Discussions

First, the performance of the GONNS is evaluated on the data generated by using a mathematical expression. Second, results of the direct implementation of the GONNS for optimization of part thickness and current combination are presented. Discussion of the results of the CoMaSA package follows.

11.5.1   Performance of the GONNS on the Simulated Systems

The surface generated by Equation (11.1) is presented in Figure 11.12. The corresponding data were used to train the neural network of GONNS. Three thousand two hundred cases were used to train the ANNs. The learning performance of the ANN of the GONNS is presented in Figure 11.13. To evaluate the performance of the modeling, a test file was prepared by selecting the points at the middle of the training data grid in Figure 11.12 and used to

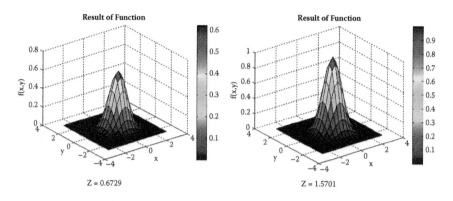

FIGURE 11.12
Theoretical surfaces of Equation (11.1). The data are used for training of ANNs.

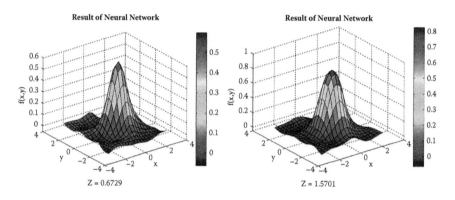

FIGURE 11.13
The ANN estimation of the surface in Figure 11.12.

test the ANN. The theoretical and ANN estimations for the test data, which were not used during the training, are presented in Figures 11.14 and 11.15, respectively. The ANN learned the characteristics of the training data with less than 2% error and the error level was the same for the test cases. The theoretical values and the estimations of the GONNS for x, y, z, and u at the peak of Equation (11.1) are presented in Table 11.1. The peak was found to have less than 20% error with respect to the range of the output. The coordinates of the peak were estimated to have less than 10% error with respect to the range of the variables. When we eliminated z from Equation (11.1), the system had two input variables. We used only 400 cases for the training. The ANNs learned the training data very quickly and the average error dropped to less than 1%. The GONNS estimated the peak and its coordinates with less than 1% error.

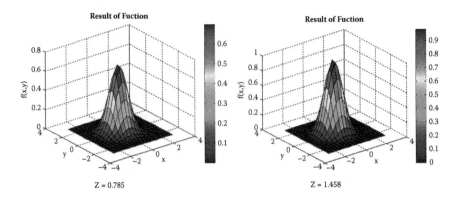

FIGURE 11.14
The test data that the ANN did not see during the training.

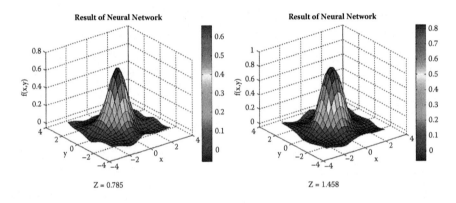

FIGURE 11.15
The ANN estimation of the test data in Figure 11.14.

TABLE 11.1

Theoretical and Estimated Coordinates and the Peak Value of Equation (11.1)

For the Peak	Values	
	Theoretical	GONNS Estimation
x	0	−0.31
y	0	0.16
z	1.57	1.459
u	1	0.831

11.5.2 Performance of the GONNS in Selecting the Optimum Composite Part Thickness and Applied Current Combination

GONNS used two ANNs in Figure 11.5 after they were trained with the semiexperimental data generated (Figures 11.6 and 11.7) by assuming that the maximum load and absorbed energy would be proportional to the thickness of plate. We anticipate a more complex relationship between the thickness and the impact resistance characteristics in practice; however, no experimental data were available for different plate thicknesses for the training. The semiexperimental data are presented in Figures 11.6 and 11.7.

The inputs of both ANNs were the plate thickness and the current. One ANN estimated the maximum load, whereas the other one estimated the absorbed energy. Back-propagation type neural networks learned the training cases with less than 0.1% error when five nodes were used at single hidden layer. A screenshot of the optimization results is presented in Figure 11.16. When the maximum absorbed energy was limited up to 120 kN and maximization of the maximum impact load was desired, the GONNS recommended a plate thickness of 4.5 mm and an application of 33.93 A.

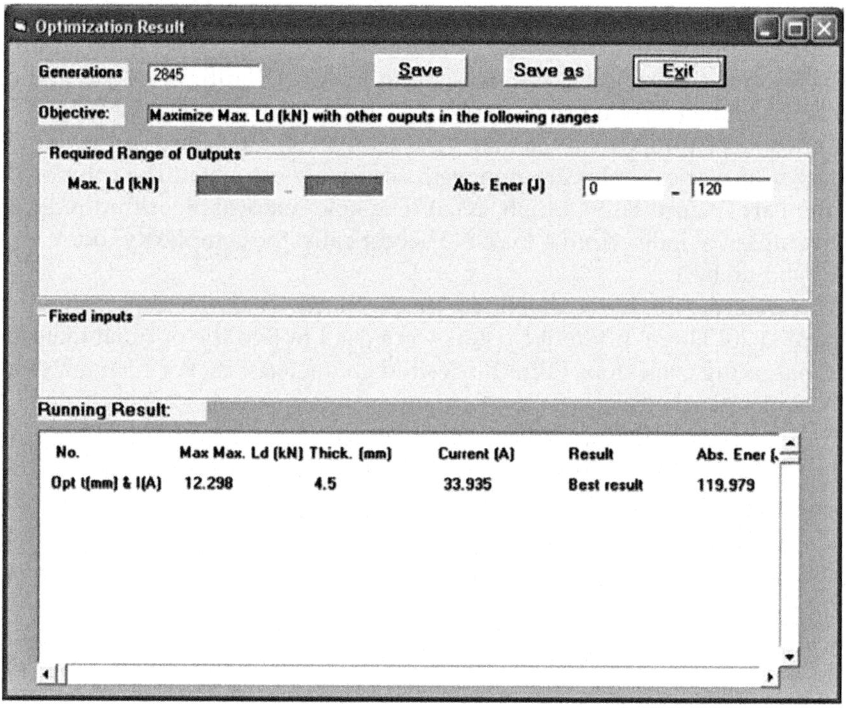

FIGURE 11.16
The optimization result screen of the user interface of the GONNS.

An experimental study is necessary to train the ANNs of the GONNS for the application of this method. It is preferable that data should be collected for five different plate thicknesses and current levels to improve the estimation accuracy. However, the current study indicates that GONNS is ready to model and find the optimal thickness and current combination for future applications.

11.5.3 Performance of the CoMaSA

The performance of the CoMaSA depends on the accuracy of the complexity estimation and GONNS. Tansel et al. (2006) and Yang et al. (2003b) studied the accuracy of the complexity index estimation by evaluating the effect of the resolution adjustment while the STL files are prepared. CAD packages use small triangles to satisfy the demand for better resolution and the increase in the number of triangles. One study indicated that when the triangles of the STL file increased from 106 to 2060, complexity changed only from 1.84 to 1.87 and was not considered significant (Yang et al., 2003b). The same group also studied the variation in the complexity of the length of parts with different profiles and curvatures. The estimations were found to be realistic and showed similar trends with respect to other composite material cost estimation packages (Butler, 1994; Eaglesham, 1998; Girivasan et al., 2000). The holes increased the estimated complexity index of the part. The relationship between the number of holes and the complexity index is presented in Figure 11.17. The radius of the holes also affected the complexity index estimation (Figure 11.18). The outlined algorithm, which calculates the complexity after the triangles on the surfaces of vertical holes are removed, accurately estimated the complexity of the part (Figure 11.19) (Singh, 2002). The hole removal algorithm reduced the complexity index from 4 to 0.98. Theoretically, the complexity index for a cube should be 1.

The diagram of the GONNS of the CoMaSA package is presented in Figure 11.20. Three different clusters were used to find the optimal material and operating conditions with the desired characteristics. For each material,

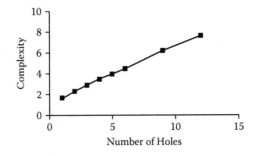

FIGURE 11.17
Increase of the complexity index with the number of holes.

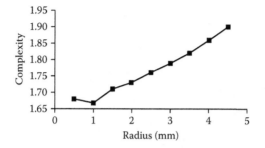

FIGURE 11.18
The size of the hole radius also increased the estimated complexity index.

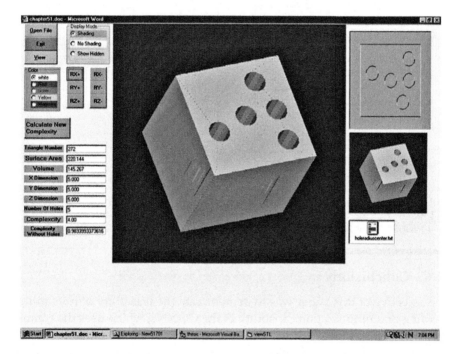

FIGURE 11.19
The user interface of the complexity estimation program of the CoMaSA complexity index decreased from 4 to 0.98 after the triangles at the surfaces of the five vertical holes were ignored during the calculation (Singh, 2002).

three ANNs estimated the cost, manufacturing time, and strength. Separate GAs searched for the best values of the inputs to optimize one of the three outputs. The user interface of the GONNS program of the CoMaSA package is presented in Figure 11.21. The optimization results are presented in Table 11.2.

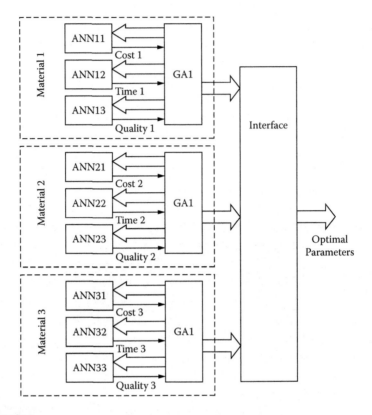

FIGURE 11.20
Block diagram of the GONNS of the CoMaSA package.

11.6   Conclusions

The objective of this paper was to demonstrate the feasibility of humanlike learning and optimization capability of the GONNS, which uses the combination of ANN and GA. The GONNS was used to model the characteristics of a mathematical expression first. Later, two composite material optimization problems were considered. Analytical, semiexperimental, and simulated data were used for the training of ANNs. The GONNS was directly implemented to look for the optimal solution for the mathematical expression and semiexperimental data of impact resistance increase by applying electricity. For cost estimation, the CoMaSA package used additional programs to calculate the part complexity index directly from the data exchange files of the commercial CAD software.

ANNs learned the characteristics of the analytical expressions with two and three inputs. When the function was written with only two inputs, ANNs learned the characteristics and the GONNS found optimal values with less

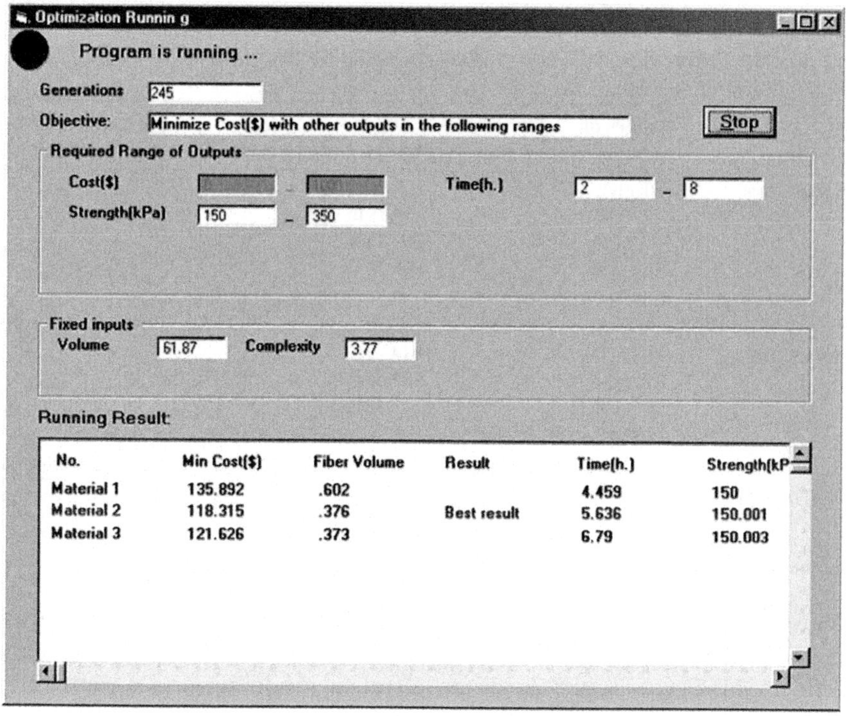

Optimization Running

Program is running ...

Generations: 245

Objective: Minimize Cost($) with other outputs in the following ranges

Stop

Required Range of Outputs

Cost($) [   ] _ [   ]   Time(h.) [2] _ [8]

Strength(kPa) [150] _ [350]

Fixed inputs

Volume [61.87] Complexity [3.77]

Running Result:

No.	Min Cost($)	Fiber Volume	Result	Time(h.)	Strength(kP
Material 1	135.892	.602		4.459	150
Material 2	118.315	.376	Best result	5.636	150.001
Material 3	121.626	.373		6.79	150.003

FIGURE 11.21
Selected optimum materials and operating conditions (minimize the cost) (Yang, 2000).

than 1% error. When we used three inputs, the average estimation accuracy of the ANN decreased to 2%, and the GONNS found the optimal values of the inputs with less than 10% error. Results indicated that the GONNS is a reliable tool as long as ANNs are properly trained. The GONNS may be used for many optimization problems of the composite materials when only the experimental data are available. The GONNS may use multiple clusters for complex problems.

The study on the optimization of the applied current and thickness combination demonstrated the feasibility of the GONNS for this application. Since we used the available experimental data selectively and generated part of the training cases with linear interpolation, additional experimental data and retraining of ANNs are necessary for future practical use. We expect a nonlinear relationship between material thickness and the two parameters that represented impact resistance (maximum load and absorbed energy) for most of the composite materials. When data are available, the ANNs could be easily trained and the GONNS could be used to determine the optimal operating conditions.

CoMaSA worked with minimum operator input and made reasonable estimations on the simulated data. The feasibility of the complexity estimation

TABLE 11.2

Results of the Optimization Performed by the GONNS Program of the CoMaSA
Package (Yang et al., 2003b; Yang, 2000)

Goal	Ranges of Output	Optimal Values	Other Outputs	Iterations	Fiber Volume	Best Material
Minimum cost	Time: 2–8 h Strength: 150–350 kPa	118.3 ($)	Time: 5.631 h Strength: 150.0 kPa	345	0.376	Material 2
Minimum manufacturing time	Cost: $100–200 Strength: 170–350 kPa	5.037 h	Cost: $143.912 Strength: 170.0 kPa	175	0.698	Material 1
Maximum strength	Cost: $50–220 Time: 3–12 h	302.669 kPa	Cost: $219.965 Time: 11.702 h	230	0.854	Material 3

programs has been previously demonstrated. Removal of the vertical holes from the part geometry was found to be very beneficial for accurate cost estimation.

The GONNS was found to be an effective modeling and optimization approach when only experimental or observed data are available for a system. Many scientific packages such as MATLAB have ANN and GA toolboxes. The GONNS approach may be easily implemented by using these packages to solve many optimization problems.

References

Boothroyd, G., and Dewhurst, P. 1991. *Product design for assembly*. Wakefield: Boothroyd Dewhurst, Inc.

Butler, A. C. 1994. Discussion of accounting theory from an engineering design and manufacturing perspective. In *Proceedings of the ASME Design for Manufacturability Conference*, Chicago, IL, March 14–16, pp. 77–87.

Carroll, D. L. 1996. Chemical laser modeling with genetic algorithms. *AIAA Journal* 34:338–346.

Chryssolouris, G., Domroese, M., and Beaulieu, P. 1992. Sensor synthesis for control of manufacturing processes. *Journal of Engineering for Industry – Transactions of the ASME* 114:158–174.

DARPA. 1988. *Neural network study*. Fairfax: AFCEA International Press.

Dayhoff, J. E. 1990. *Neural network architectures: an introduction*. New York: Van Nostrand Reinhold.

Eaglesham, M. A. 1998. A decision support system for advanced composites manufacturing cost estimation. Ph.D. thesis, Virginia Polytechnic Institute and State University.

Girivasan, V., Yang, S. Y., Kropas-Hughes, C. V., Tansel, I. N., Sasirathsiri, A., and Bao, W. Y. 2000. Automated manufacturing time and cost estimation of composite parts using neural networks. In *Smart engineering systems: neural networks, fuzzy logic, evolutionary programming, data mining, and complex systems*, ed. C. H. Dagli, A. Buczak, J. Ghosh, M. J. Embrechts, O., Ersoy, and S. Kercel, Vol. 10, 989–994. New York: ASME Press.

Goldberg, D. E. 1989. *Genetic algorithms in search, optimization and machine learning.* Reading, MA: Addison-Wesley.

Gutowski, T., Hoult, D., Dillon, G., Neoh, E., Muter, S., and Tse, M. 1994. Development of a theoretical cost model for advanced composite fabrication. *Composites Manufacturing* 5:231–239.

Hecht-Nielsen, R. 1990. *Neurocomputing.* Reading, MA: Addison-Wesley Publishing Company.

Hess, R. W., and Romanoff, H. P. 1987. *Aircraft airframe cost estimating relationships, Rept. R-3255-AF.* Santo Monica, CA: Rand Corp.

Kaplan, R. S., and Atkinson, A. A. 1989. *Advanced management accounting,* 2nd ed. Englewood Cliffs: Prentice-Hall.

Kim, C. E. 1993. Composite cost modeling: complexity. M.S. thesis, Massachusetts Institute of Technology.

Ko, T. J., and Kim, H. S. 1998. Autonomous cutting parameter regulation using adaptive modeling and genetic algorithms. *Precision Engineering* 22:243–251.

Krebs, J., Bhattacharyya, D., and Friedrich, K. 1997. Production and evaluation of secondary composite aircraft components—a comprehensive case study. *Composites Part A* 28A:481–489.

Ladd, R. S. 1996. *Genetic algorithms in C++.* New York, NY: M&T Books.

LeBlanc, D. J. 1976. Advanced composites cost estimation manual, Vol. 1 AFFDL-TR-76-87, Northrop Corp.

Li, M., Kendall, E., and Kumar, J. 1997. A computer system for lifecycle cost estimation and manufacturability assessment of composites. *Proceedings of the ICCM-11*, Australia.

Mabson, G. E., Flynn, B. W., Ilcewicz, L. B., and Graesser, L. D. 1994. The use of COSTADE in developing composite commercial aircraft fuselage structures. *Proceedings of 35th AIAA/ASME/ASCE/AHS/ASC Conf. Hilton Head*, C, April 18–20, Part 3.

Masters, T. 1995. *Advanced algorithms for neural networks.* New York: John Wiley & Sons, Inc.

Muter, S. 1993. Cost comparison of alternate designs: an information based model. M.S. thesis, Massachusetts Institute of Technology.

Niazi, A., Dai, J. S., Balabani, S., and Seneviratne, L. 2006. Product cost estimation: technique classification and methodology review. *Journal Manufacturing Science Engineering – Transactions of ASME* 128:563–576.

Pearce, D. 1989. A statistical/heuristic approach to estimating molds. Annual Technical Conference—Society of Plastics Engineers. Publ. by Soc of Plastics Engineers, Brookfield Center, CT, USA, pp. 364–366.

Pugh, S. 1991. *Total design: integrated methods for successful engineering.* Wokingham, UK: Addison Wesley.

Ramkumar, R. L., Vastava, R. B., and Saha, S. K. 1991. Manufacturing cost model for composites. *Proceedings of 23rd International SAMPE Technical Conference*, pp. 982–994.

Rohani, M., and Dean, E. B. 1996. Toward manufacturing and cost consideration in multidisciplinary aircraft design. 37th AIAA/ASME/ASCE/AHS/ASC Structures, Structural Dynamics and Materials Conference, Salt Lake City, UT, April 15–17, WIP Paper no. AIAA-96-1620-CP, Part 4, 2602–2612.

Rumelhart, D. E., Hilton, G., and Williams, R. J. 1986. Learning internal representations by error propagation. *Parallel distributed processing: explorations in the microstructure of cognition*, Vol.1, ed. E. Rumelhart and J. L. McClelland. Cambridge, MA: Massachusetts Institute of Technology Press.

Sierakowski, R. L., Telitchev, I. Y., and Zhupanska, O. I. 2008. On the impact response of electrified carbon fiber polymer matrix composites: effects of electric current intensity and duration. *Composites Science and Technology* 68:639–649.

Singh, R. N. 2002. Improvement of existing complexity estimation method. M.S. thesis, Florida International University.

Tansel, I. N., 1990. Neural network approach for representation and simulation of 3D-cutting dynamics. *Transactions of North American Metal Research Institute*, pp. 193–200.

Tansel, I., Bao, W. Y., Arkan, T. T., and Tansel, B. 1997. Visualization of underground contamination by using neural networks. In *Smart engineering systems: neural networks, fuzzy logic, evolutionary programming, data mining, and rough sets*, ed. C. H. Dagli, M. Akay, O. Ersoy, B. R. Fernandez, and A. Smith. New York, NY: ASME Press.

Tansel, I., Yang, S. Y., Shu, C., Bao, W. Y., and Mahendrakar, N. 1999. Introduction to genetically optimized neural network systems (GONNS). In *Smart engineering systems: neural networks, fuzzy logic, evolutionary programming, data mining, and rough sets*, ed. C. H. Dagli, A. Buczak, O. Ersoy, and B. R. Fernandez, 331–336. New York, NY: ASME Press.

Tansel, I. N., Ozcelik, B., Bao, W. Y., Chen, P., Rincon, D., Yang, S. Y., and Yenilmez, A. 2006. Selection of optimal cutting conditions by using GONNS. *International Journal of Machine Tools & Manufacture* 46:26–35.

Tse, M. 1990. Design cost model for advanced composite structures. M.S. thesis, Massachusetts Institute of Technology.

Venkataraman, G. 2000. Development of a composite material selection advisor for polymer matrix composites (PMCs). M.S. thesis, Florida International University.

Winter, G., Cuesta, P., Periaux, J., and Galan, M. 1996. *Genetic algorithm in engineering and computer science*. New York, NY: John Wiley & Sons.

Wong, J. P., Imam, I. N., Khosravi-Kamrani, A., Parsaes, H. R., and Tayyari, F. 1992. A totally integrated manufacturing cost estimating system (TIMCES). In *Economics of advanced manufacturing systems*, ed. H. R. Parsei and A. Mital, 201–224. London: Chapman & Hall.

Yang, S. 2000. Genetically optimized complex neural network system. M.S. thesis, Florida International University.

Yang, S. Y., Girivasan, V., Singh, N. R., Tansel, I. N., and Kropas-Hughes, C. V. 2003a. Selection of optimal material and operating conditions in composite manufacturing: I. Computational tool. *International Journal of Machine Tools & Manufacture* 43:169–173.

Yang, S. Y., Girivasan, V., Singh, N. R., Tansel, I. N., and Kropas-Hughes, C. V. 2003b. Selection of optimal material and operating conditions in composite manufacturing: II. Complexity, representation of characteristics and decision making. *International Journal of Machine Tools & Manufacture* 43:175–184.

12

Development of a Prototype Computational Framework for Selection of Natural Fiber-Reinforced Polymer Composite Materials Using Neural Network

S. M. Sapuan and I. M. Mujtaba

CONTENTS

ABSTRACT A development of a neural network–based materials selection system for natural fiber polymer composites is presented. Datasets for input and output parameters are collected from published papers, which are composed of datasets for tensile strength and modulus (inputs) as well as flexural strength (output). The material selection for horizontal shelf is chosen as a case study. The materials selection process involves two stages. In the first stage, the correlation between the inputs and the output is predicted and materials whose output properties are very close to the predicted output property are selected. In the second stage of the selection, a method called multiattribute ranking is used to decide which is the most suitable material.

12.1   Introduction

The selection of composite materials, compared to homogeneous materials such as metals and plastics, is quite difficult to perform because of the anisotropic nature of the materials requiring tailor-made consideration of products and materials. Despite this obstacle, various computerized composite materials selection systems were developed by some experts using materials database (Baur, 1995) and knowledge-based systems (Sapuan et al., 2002; Sapuan and Abdalla, 1998).

In this paper, the development of a materials selection system for natural fiber-reinforced polymer composites is presented using neural network (NN). NN or artificial neural network (ANN) is a computer system that simulates the neurons of a biological nervous system or computational models of the brain. In the past, various research works have been carried out to develop materials selection system using ANNs. Goel and Chen (1996) conducted research on materials selection using an expert system and NN. They called their system expert network. In their work, knowledge about the materials was stored in an NN, and the output of the NN was validated using this expert system. The NN was used to select the optimum materials from the database. The materials stored in the system were mainly metals. The work of Cherian et al. (2001) was concerned with the development of an NN material selection system for powder metallurgy. Inputs to the system were the mechanical properties of the powder metallurgy materials such as tensile strength, hardness, and percentage of elongation, and the predicted outputs were metal powder compositions and process setting. Similar work was also conducted by Smith et al. (2002), but this time they used the inverse problem of NN in the selection of powder metallurgy materials. In their work, the inputs were final properties and the outputs were the powder metallurgy processing parameters.

So far, only a few publications have reported on the application of NNs in composite material field (Zhang and Friedrich, 2003), and very limited literature has been found in the area of materials selection for composite materials. Recent papers published in *Materials and Design* on NNs in the area of composite materials (Altinkok and Koker, 2004, 2006) were found to be suffering from various discrepancies as pointed out by Sha and Edwards (2007), proving that general understanding of the research area has not yet been established and considerable work remains to be carried out in this particular field. Yang et al. (2003a, 2003b) attempted to use NN in the selection of optimum materials in composite manufacturing; NN was used in conjunction with other software tools such as genetic algorithm and computer-aided design/computer-aided manufacturing. However, none of the cited research has looked into the study of NNs in the selection of natural fiber-reinforced composites, and therefore, in light of this background, this work is presented.

12.2 Data Collection

In natural fiber-reinforced polymer composite selection, the main problem in obtaining reliable experimental data to be used in materials selection with NN is the large combination of parameters in developing composite materials. As shown in Figure 12.1, there are many ways to create polymer composites comprising various types of fibers and fillers, a wide variety of natural fibers available, and with different types of matrix from the family of thermosetting and thermoplastic polymers.

For the purpose of this research, the data were gathered from the previous published work of the principal author and his coworkers carried out at Universiti Putra Malaysia (Sastra et al., 2005, 2006; Bachtiar et al., 2008; Maleque et al., 2007; Sapuan et al., 2003, 2006; Arib et al., 2006a, 2006b; Lai et al., 2005). In addition, data from various relevant articles were also collected in order to enrich the database (Joseph et al., 2002; Wambua et al., 2003; Idicula et al., 2005; Dhakal et al., 2007; Tan, 1997; Jayaraman and Battacharyya, 2004; Herrera-Franco and Valadez-Gonzalez, 2004). A total of 121 datasets, which represent 121 different types of natural fiber composites (different combination of fiber weight, treatment method, fiber types, and other parameters) were collected for optimum NN construction. For each

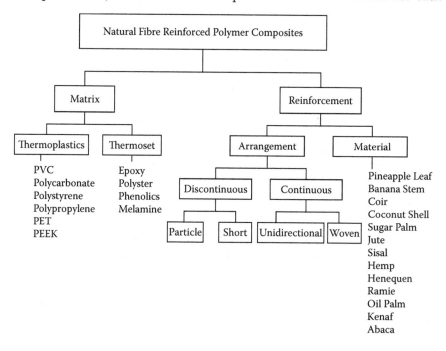

FIGURE 12.1
Complex structural components of natural fiber-reinforced polymer composites.

material, the properties in the selected datasets were tensile strength, tensile modulus, and flexural strength. The total datasets (S) are represented as:

$$S = [(I^{(i)}, O^{(i)})|\ i = 1, ..., P] \tag{12.1}$$

where P is the maximum number of dataset. These total datasets were divided into training datasets (R), testing datasets (T), and validation datasets (V). The training datasets are

$$R = [(I^{(i)}, O^{(i)})|\ i = 1, ..., M] \tag{12.2}$$

where M is the maximum number of training datasets. The testing datasets are

$$T = [(I^{(i)}, O^{(i)})|\ i = M + 1, ..., N] \tag{12.3}$$

where N is the maximum number of testing datasets. The validation datasets are

$$V = [(I^{(i)}, O^{(i)})|\ i = N + 1, ..., P] \tag{12.4}$$

where $I^{(i)}$ is the ith input parameter and $O^{(i)}$ is the ith output parameter.

In this study, approximately 50% of the total datasets are for training datasets ($M = 61$ datasets), 30% for testing ($N - M = 30$ datasets), and 30% for validation ($P - N = 30$ datasets). Selected datasets used are tabulated in Table 12.1.

There are various challenges in obtaining comparable data of natural fiber composites such as:

- The problems in obtaining complete sets of data on tensile strength, tensile modulus, and flexural strength (many papers were excluded due the incomplete sets of data).
- Fiber fractions were given both in weight and volume, and it is not very clear whether the researchers were consistent in their definitions.
- Insufficient information—for example, no mention about the fiber fraction, the length of fiber (for short fiber), etc.

As a result, a large variation was observed in the data; for instance, the minimum value of tensile modulus was 337 MPa for coir/polypropylene composite and 11,590 MPa for jute/polypropylene composite (see Table 12.1). This caused the distribution to be biased toward the lower end value of the data. To minimize this variation as much possible, data were taken from the previous work of the principal author.

TABLE 12.1

Selected Datasets Used in the Study

Dataset No.	Fiber Type	Matrix Type	Typical Parameters Studied	Tensile Strength (MPa)	Tensile Modulus (MPa)	Flexural Strength (MPa)
1	Sugar palm	Epoxy	Untreated short random 10% w.w. fiber	32	1081	73.3
2	Sugar palm	Epoxy	NaOH treated at 0.25 M concentration and 1 h soaking time 10% w.w. fiber	49.9	3782	96.7
3	Banana stem	Epoxy	Untreated woven fiber	46	1890	74
4	Banana stem	Epoxy	Untreated woven fiber	14	976	26
5	Pineapple leaf	Polypropylene	Untreated unidirectional 2.7% by volume fiber	36	550	69.5
6	Coir	Polypropylene	Stearic acid treated short (0.315 mm length) 15% w.w. fiber	24	337	1.9
7	Coconut shell particle	Epoxy	Untreated 10% w.w. filler particle	30	2780	67
8	Banana stem	Phenol formaldehyde	Untreated 10 mm length 45% by volume fiber	12	268	25
9	Kenaf	Polypropylene	Untreated short random 30% w.w. fiber	27	6800	28
10	Banana/ sisal	Polyester	Untreated short random 19% by volume fiber	39	1347	48
11	Hemp	Unsaturated polyester	Wet short random 21% by volume fiber	46	620	767
12	Jute	Polypropylene	NaOH treated 40% w.w. fiber	152	11,590	74
13	Wood	HDPE waste	Wet short random 20% w.w. fiber	21.2	2135	27.4
14	Henequen	HDPE	Silane-treated unidirectional longitudinal fiber	72	2600	97

Sources: Sastra et al. (2005, 2006), Bachtiar et al. (2008), Maleque et al. (2007), Sapuan et al. (2003, 2006), Arib et al. (2006a, 2006b), Lai et al. (2005), Joseph et al. (2002), Wambua et al. (2003), Idicula et al. (2005), Dhakal et al. (2007), Tan (1997), Jayaraman and Battacharyya (2004), Herrera-Franco and Valadez-Gonzalez (2004).

The datasets (inputs and output) were stored in MS Excel and were normalized in the range of -1 and $+1$ using the following formula to facilitate data training, testing, and validation:

$$X_n = [2(X - X_{min})/(X_{max} - X_{min})] - 1 \qquad (12.5)$$

where X_n is the normalized value of the parameters, and $X_{max} - X_{min}$ are the minimum and maximum values of variable X, respectively.

12.3   Determination of Product and Properties

In the following subsections, product selection, material selection requirements, and system architecture are presented.

12.3.1   Product Selection

Natural fiber-reinforced composites comprise a complicated combination of different parameters. To reduce this complication, the selection of very simple components for the materials selection process is considered a wise decision. In addition, it is known from the literature (Sastra et al., 2005, 2006; Bachtiar et al., 2008; Maleque et al., 2007; Sapuan et al., 2003, 2006; Arib et al., 2006a, 2006b; Lai et al., 2005; Joseph et al., 2002; Wambua et al., 2003; Idicula et al., 2005; Dhakal et al., 2007; Tan, 1997; Jayaraman and Battacharyya, 2004; Herrera-Franco and Valadez-Gonzalez, 2004) that the application of natural fiber is currently restricted to nonstructural or, at most, semistructural components, where only significantly small amounts of load-bearing capacity is anticipated. For this purpose, it was decided to use a horizontal multipurpose shelf, as shown in Figure 12.2, as the item upon which the materials selection process will be performed.

The dimensions of the shelf are as follows: length = 1220 mm, thickness = 15 mm, width = 155 mm. The distance from the end to the support is 260 mm.

12.3.2   Material Selection Requirements

It was obvious from the data collection stage (Section 2.2) that strength and modulus are the two important properties in the selection process for structural components. In fact, strength and stiffness are often referred to as the materials selection drivers (Sapuan and Abdalla, 1998). Tensile strength is normally regarded as a very important property, but for horizontal shelves, where the load is predominantly bending in nature, the flexural strength of the materials becomes significant and will therefore be the output in the study. In addition, there is only one output parameter that is fixed in this

FIGURE 12.2
A multipurpose wooden horizontal shelf.

study because, as reported by Zhang and Friedrich (2003), majority of the work on NNs in composite materials have only one output, and in order to be consistent, this norm is followed. When load is applied to the shelf, the shelf has to sustain the constant load applied at a prolonged time and in such a situation, the potential for creep failure is significant. Thus, the study considered such potential failure.

The parameters consist of numerical (tensile strength and modulus, and flexural strength) and qualitative (aesthetics, manufacturability, availability, and cost) data. In this study, it was difficult to obtain comparable creep data for all the materials and as a consequence, they were not included in the materials selection system. All input and output parameters are broadly divided into three main categories to satisfy the materials selection requirements: functional, manufacturing, and economic considerations (Figure 12.3).

As far as manufacturing requirement is concerned, the aspect of manufacturability and aesthetics are selected. Manufacturability refers to the ease of accomplishing a certain manufacturing process to produce a particular material. If the manufacturing process is a manual lay-up process, the process is straightforward and the cost to produce the mold and the sample is affordable; therefore, it is given high rating. As for the injection molding process, the cost of machinery is high and the production of specimens is more complex, thereby earning the process a lower rating based on a scale of 1 to 5. Nowadays, product design should consider both the engineering design aspect as well as the industrial design aspect of a product. Issues such as form, shape, color, texture, ergonomics, and

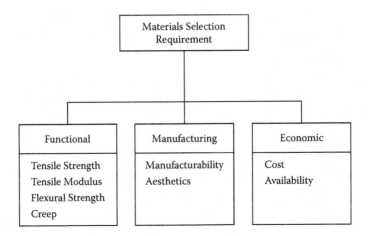

FIGURE 12.3
Materials selection requirements for natural fiber composite shelf.

aesthetics must also be given attention. In line with this call, aesthetics of the product is selected in this study. It is categorized under the manufacturing requirement because the aesthetics of a product is ensured during the manufacturing stage.

In the economic requirement, cost and availability are regarded as important. Direct material cost is very important in materials selection, but to date, there is no published work on cost comparison of natural fiber composites. Some of the materials can be obtained at no cost at all and they are regarded as agricultural waste. Therefore, only qualitative data can be provided. Availability can be regarded as indirect cost because, if the material is cheap but is not locally available, the cost to import the material can be high. Its in-house unavailability incurs high cost. Once again, the actual data of such an item are not available and only qualitative data can be provided.

12.3.3 System Architecture

The general structure of the computer-aided materials selection system using NN is given in Figure 12.4, where the relationship between various tools used in the materials selection, including the NN, materials database, product design tools, and materials data sources are shown. The data were stored in an MS Excel spreadsheet and the NN was developed using the MATLAB® neural network toolbox. Figure 12.5 shows the detailed architecture of the materials selection system using NN, which can be divided into two stages of materials selection process.

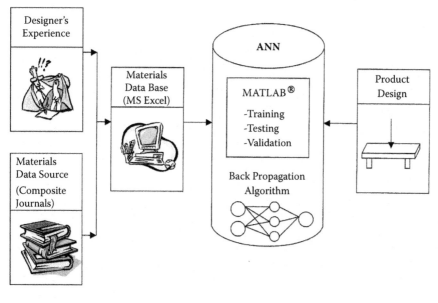

FIGURE 12.4
Structure of the materials selection system.

12.4 First Stage of Material Selection

The MATLAB® neural network toolbox was used in the development of a material selection system for natural fiber-reinforced polymer composites. NN modeling is a nonlinear statistical analysis method that is in the form of a "black box." In this black box, the input data and output data are connected via a set of nonlinear functions. In this paper, a fully connected four-layer (one input, two hidden, and one output layers) feedforward network with sigmoid transfer functions is used (Figure 12.6).

12.4.1 Data Training

Before the NN composite material selection system can be applied, the procedures for obtaining the NN model, that is, the forward model used in these strategies, are initially performed together with the method of training the system.

12.4.2 Forward Modeling

The procedure of training an NN to represent the dynamics of the system is referred to as forward modeling. Forward modeling, in this case, refers to training the NN model to predict the output of property at the next instant of time $(t + 1)$. Data used for the development of the model are shown in Table 12.2.

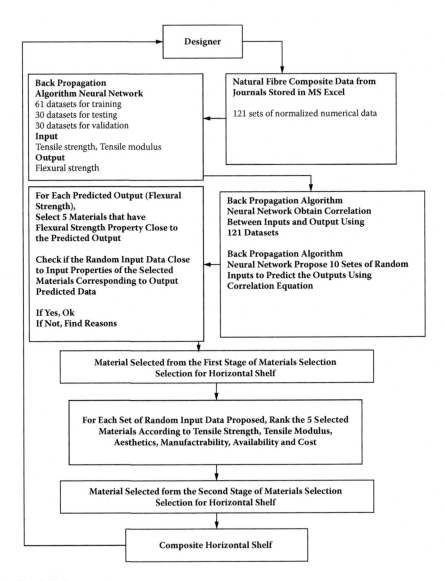

FIGURE 12.5
Detailed architecture of the materials selection system using neural network.

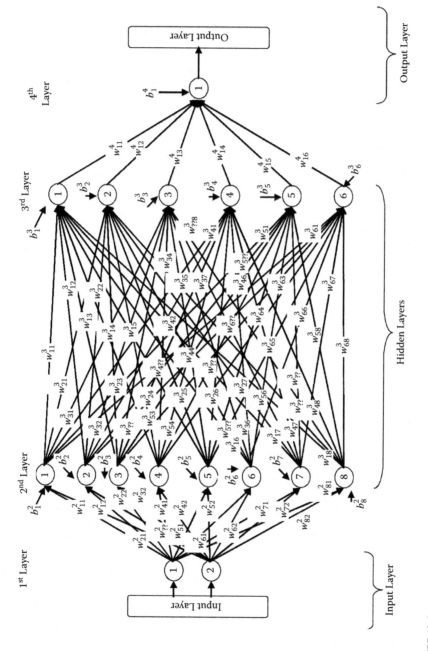

FIGURE 12.6
A typical structure of fully connected four-layer network.

TABLE 12.2

Data Used for Development of the Model (Italic Data Denote Validation and Bold Data Denote Test Dataset)

Input1	Input2	Output	Prediction	Input1	Input2	Output	Prediction	Input1	Input2	Output	Prediction
19	370	1.67	23	350	2	20	291	34	38	600	69.5
21	365	1.73	27.5	325	2.05	25	3500	34	41.4	3751	72.6
26	*323*	*1.75*	*27*	*330*	*2.1*	*16*	*370*	*36*	*148*	*11,450*	*73*
23	**332**	**1.77**	**24**	**337**	**2.13**	**30**	**4141**	**38**	**32**	**1081**	**73.3**
21.5	380	1.77	27	350	2.13	31	600	43	46	1890	74
20	375	1.78	27	328	2.15	126	7860	43	152	11,590	74
22	*352*	*1.8*	*24*	*330*	*2.15*	*20*	*375*	*45*	*46*	*620*	*76*
20	**355**	**1.8**	**27**	**340**	**2.17**	**128**	**7920**	**47**	**30.8**	**1032**	**77.5**
21	365	1.8	30	322	2.25	39	1347	48	38	620	79
22	375	1.8	2	550	3	23	398	50	17	560	80
21.5	*363*	*1.82*	*7*	*175*	*10*	*26*	*556*	*50*	*35*	*2750*	*81*
22	**357**	**1.83**	**13.2**	**559**	**11.3**	**25**	**1400**	**51**	**172**	**12,150**	**82**
24	340	1.85	12.9	1026	13.9	51	1443	53	52	1000	83
21	*370*	*1.85*	*15.4*	*1379*	*17.6*	*52*	*6800*	*54*	*32.5*	*1153*	*83.3*
22	*350*	*1.87*	*3*	*600*	*19*	*58*	*1597*	*56*	*44.3*	*1161*	*84*
24	**352**	**1.87**	**94**	**5250**	**19**	**27**	**2300**	**56**	**30.64**	**3657**	**85.3**
24	348	1.88	19.9	2356	21.8	142	8390	56	41.85	3772	90.7
26	*332*	*1.9*	*101*	5280	22	23	720	57	49.6	1196	92.6
24	*337*	*1.9*	*24.4*	*1483*	*23*	*37.6*	*3849*	*57.1*	*52.3*	*1220*	*93*
23	**338**	**1.9**	**22.8**	**1282**	**24.4**	**29**	**640**	**60**	**63**	**680**	**94**
25	345	1.9	21.2	2049	24.7	57	1601	62	49.9	3782	96.7
26	340	1.95	12	268	25	37.9	3869	64.4	72	2600	97
24	*368*	*1.96*	*14*	*976*	*26*	*30.5*	*1056*	*64.7*	*51.7*	*1255*	*108*

25	360	1.97	33	5300	26	48.4	5250	64.9	56	1270	113
22.5	350	1.98	22.7	2410	26.6	32	620	65.6	81	2750	131
22	358	1.98	10	1300	27	35	540	66.2			
27	322	1.99	23.1	2135	27.4	42.7	3247	66.5			
26	330	1.99	20.6	2400	27.7	30	2780	67			
25	342	1.99	114	5740	28	43	1664	67.7			
27	328	2	27	6800	28	39	660	68			
24.5	330	2	27.1	2711	32	40.7	3517	69.4			
22.5	348	2	6	197	34	36	550	69.5			

The correlation can be found by expanding Equation (12.1) given by Tanvir and Mujtaba (2006). An architecture similar to the one used in this work is shown in Figure 12.7; weights and biases are shown in Tables 12.3, 12.4, and 12.5, and their scale-up parameters in Table 12.6. Figure 12.8 shows the MATLAB output graph for the dataset used.

$$a_1^4 = f_1^4 \left(\sum_{k=1}^{4} w_{1k}^4 \times a_k^3 + b_1^4 \right) \tag{12.6}$$

All symbols used here are similar to those used by Tanvir and Mujtaba (2006).

Table 12.7 shows the results of the predicted output data after 10 sets of random data (designer's choice) are used as input into the correlation equation.

The designer proposed a set of inputs and by using the ANN, an output is predicted. In this first stage of selection, five materials whose flexural strength values are very close to the predicted output are selected as the candidate materials. In this study, the designer proposed 10 sets of input data. The results of materials selection based on all 10 predicted data outputs (five materials for each output) are shown in Table 12.8.

For the random inputs of 90 and 7500 MPa for tensile strength and tensile modulus, respectively (random input dataset 1), the output (flexural strength) predicted from the correlation is 50.055 MPa. Five materials with values ranging from 48 to 53 MPa are selected. Next, the data for original tensile strength and tensile modulus are obtained for each material.

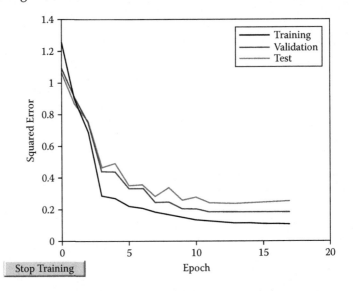

FIGURE 12.7
MATLAB training graph for the data used above.

TABLE 12.3

Weights and Bias Between Input Layer and First Hidden Layer

w_{11}^{iw}	0.4028	w_{12}^{iw}	1.4924	b_1^{iw}	−7.3513
w_{21}^{iw}	−1.4197	w_{22}^{iw}	2.9759	b_2^{iw}	1.8018
w_{31}^{iw}	−3.1381	w_{32}^{iw}	−0.08906	b_3^{iw}	−0.02609
w_{41}^{iw}	0.74888	w_{42}^{iw}	−1.4314	b_4^{iw}	−0.01731
w_{51}^{iw}	−0.77206	w_{52}^{iw}	−1.8204	b_5^{iw}	3.099
w_{61}^{iw}	1.3506	w_{62}^{iw}	0.8047	b_6^{iw}	−1.3228
w_{71}^{iw}	−1.1845	w_{72}^{iw}	1.2042	b_7^{iw}	−3.336
w_{81}^{iw}	0.014626	w_{82}^{iw}	−1.6894	b_8^{iw}	6.5478

TABLE 12.4

Weights Between the First and Second Hidden Layers

w_{11}^{lw1}	0.21316	w_{12}^{lw1}	0.002401	w_{13}^{lw1}	−0.19579	w_{14}^{lw1}	0.60429
w_{21}^{lw1}	0.48394	w_{22}^{lw1}	−2.031	w_{23}^{lw1}	2.134	w_{24}^{lw1}	−0.45785
w_{31}^{lw1}	0.18728	w_{32}^{lw1}	−0.33503	w_{33}^{lw1}	0.7647	w_{34}^{lw1}	0.79053
w_{15}^{lw1}	1.6213	w_{16}^{lw1}	0.10797	w_{16}^{lw1}	0.69935	w_{16}^{lw1}	−0.74987
w_{25}^{lw1}	0.89027	w_{26}^{lw1}	−0.58913	w_{26}^{lw1}	0.37767	w_{26}^{lw1}	−0.93146
w_{35}^{lw1}	−0.91309	w_{36}^{lw1}	−1.5791	w_{36}^{lw1}	−0.7507	w_{36}^{lw1}	0.47401
b_1^{lw1}	1.4149	b_2^{lw1}	−0.29962	b_3^{lw1}	1.5828		

TABLE 12.5

Weights Between the Second Hidden Layer and the Output Layer

1.1153	−2.1673	−1.5048	−0.6006

TABLE 12.6

Scale-up Parameters

std_P1	30.377	mean_P1	37.191
std_P2	2419.5	mean_P2	1817.2
std_T	33.797	mean_T	37.389

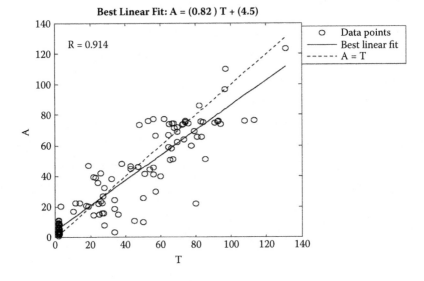

FIGURE 12.8
MATLAB output graph for the dataset used.

TABLE 12.7

NN Prediction for the Input Provided

Input1 (tensile strength, MPa)	Input 2 (tensile modulus, MPa)	Prediction (flexural modulus, MPa)
90	7500	50.055
15	12,000	78.347
35	5700	30.349
50	1500	75.937
4	350	21.24
100	800	92.475
150	6000	16.513
18	390	13.896
20	3500	6.6588
77	3850	132.84

TABLE 12.8

Results of Materials Selection Based on All 10 Predicted Data Outputs (Five Materials for Each Output)

Output predicted (MPa)	50.055	78.347	30.349	75.937	21.24	92.475	16.513	13.896	6.6588	132.84
Material 1 (MPa)	51	77.5	28	73.3	19	92.6	19	10	2.25	108
Material 2 (MPa)	50	81	28	74	22	96.7	13.9	19	2.17	96.7
Material 3 (MPa)	50	79	27.4	76	23	90.7	17.6	11.3	2.15	113
Material 4 (MPa)	48	76	32	73	21.8	93	21.8	13.9	10	94
Material 5 (MPa)	53	80	27.7	74	19	94	19	17.6	3	131

Table 12.9 shows the results of correlation between random data input and predicted output as well as the flexural strength of candidate materials and their corresponding tensile strength and tensile modulus.

12.5 Second Stage of Selection

The second stage of selection is quite straightforward compared to the first one. In this stage, parameters and properties (e.g., aesthetics, manufacturability, availability, cost) are used in conjunction with flexural strength, tensile strength, and tensile modulus to obtain the most suitable material. Since it is not easy to quantify the data on aesthetics, manufacturability, availability, and cost, as explained in an earlier section, only qualitative data are used in the selection process. The data for flexural strength, tensile strength, and tensile modulus are provided numerically.

According to Sergent (1991), the material selection problem is considered trivial if there is only one property to be considered and the materials are ranked according to the value of the said property. For the case of multiattribute problems, as in the current study, various methods are available to solve them such as Simpson's paradox, Condorcet's paradox, noninferior set, Arrow's impossibility theorem, and von Neumann and Morgenstern axioms (Sergent, 1991). Arrow's impossibility theorem is used in this study. It uses the principle based on the ranked orders for seven different attributes and properties. The selection process is carried out by listing those seven properties and attributes in the top row of a table. Candidate materials for each designer input data are listed in the left column of the table. For each material, a total score is given by summing the score of each property. The candidate material with the highest score is selected. Table 12.10 shows the multiattribute ranking for the materials selected after the first stage of materials selection, resulting from the first sets of random input data proposed by the designer. A similar approach is used for the other nine random input datasets. In this study, since mixed data are available (qualitative and numerical), all data are normalized for the sake of consistency and obtaining comparable data, and are scored on the scale of 1 to 5. The desirable attributes are assigned high scores, and vice versa.

TABLE 12.9

Correlation between Random Data Input and Predicted Output, Flexural Strength of Candidate Materials, and Corresponding Tensile Strength and Tensile Modulus

Random Input Tensile Strength (MPa)	Random Input Tensile Modulus (MPa)	Predicted Flexural Strength (MPa)	Selected Flexural Strength (MPa)	Corresponding Tensile Strength (MPa)	Corresponding Tensile Modulus (MPa)
90	7500	50.055	51	25	1400
			50	26	556
			50	23	398
			48	39	1347
			53	51	1443
15	12,000	78.347	77.5	30.8	1032
			81	36	2057
			79	38	620
			76	46	620
			80	17	560
35	5700	30.349	28	27	6800
			28	114	5740
			27.4	23.1	2135
			32	27.1	2711
			27.7	20.6	2400
50	1500	75.937	73.3	32	1081
			74	46	1890
			76	46	620
			73	148	11,450
			74	152	11,590
4	350	21.24	19	94	5250
			22	101	5280
			23	24.4	1483
			21.8	19.9	2356
			19	3	600
100	800	92.475	92.6	49.6	1196
			96.7	49.9	3782
			90.7	41.85	3772
			93	52.3	1220
			94	63	680
150	6000	16.513	19	94	5250
			13.9	12.9	1026
			17.6	15.4	1379
			21.8	19.9	2356
			19	3	600
18	390	13.896	10	7	175

—continued

TABLE 12.9 (CONTINUED)

Correlation between Random Data Input and Predicted Output, Flexural Strength of Candidate Materials, and Corresponding Tensile Strength and Tensile Modulus

Random Input Tensile Strength (MPa)	Random Input Tensile Modulus (MPa)	Predicted Flexural Strength (MPa)	Selected Flexural Strength (MPa)	Corresponding Tensile Strength (MPa)	Corresponding Tensile Modulus (MPa)
			19	94	5250
			11.3	13.2	559
			13.9	12.9	1026
			17.6	15.4	1379
20	3500	6.6588	2.25	30	322
			2.17	27	340
			2.15	24	330
			10	7	175
			3	2	550
77	3850	132.84	108	51.7	1255
			96.7	49.9	2782
			113	56	1270
			94	63	680
			131	81	2750

12.6 Conclusions

A prototype computational framework for materials selection of natural fiber-reinforced polymer composites for a horizontal shelf using ANN is presented in this paper. The developed system enables designers to select appropriate natural fiber-reinforced polymer composites for horizontal shelf based on a set of predefined criteria. Ten random input datasets were proposed by the designers in the first stage of the selection process, and through the correlation equation, the output property is predicted. The materials that have the output properties close to the predicted properties are selected as candidate materials. The selection of the final optimum material is carried out using multiattribute ranking method.

TABLE 12.10

Multiattribute Ranking of the Materials for Random Input Dataset 1

Candidate Material	Tensile Strength	Tensile Modulus	Flexural Strength	Aesthetics	Manufacturability	Availability	Cost	Total Score
1	2	4	4	2	4	1	1	18
2	3	2	3	1	5	4	4	22
3	1	1	2	4	1	2	2	13
4	4	3	1	5	3	3	3	22
5	5	5	5	3	2	5	5	30

References

Altinkok, N., and Koker, R. 2004. Neural network approach to prediction of bending strength and hardening behaviour of particulate reinforced (Al-Si-Mg) aluminium matrix composites. *Materials and Design* 25:595–602.

Altinkok, N., and Koker, R. 2006. Modelling of the prediction of tensile and density properties in particle reinforced metal matrix composites by using neural networks. *Materials and Design* 27:625–631.

Arib, R. M. N., Sapuan, S. M., Hamdan, M. A. M. M., Paridah, M. T., and Zaman, H. M. D. K. 2006a. Impact and bending properties of pineapple leaf fibre (PALF) reinforced polypropylene (PP) laminated composites. *Brunei Darus Salam Journal of Technology and Commerce* 4:130–135.

Arib, R. M. N., Sapuan, S. M., Hamdan, M. A. M. M., Paridah, M. T., and Zaman, H. M. D. K. 2006b. Mechanical properties of pineapple leaf fibre (PALF) reinforced polypropylene composites. *Materials and Design* 27:391–396.

Bachtiar, D., Sapuan, S. M., and Hamdan, M. M. (2008). The effect of alkaline treatment on tensile properties of sugar palm fibre reinforced epoxy composites. *Materials and Design* 29:1285–1290.

Baur, E. 1995. English version of FUNDUS material database. *Reinforced Plastics* 39:21.

Cherian, R. P., Smith, L. N., and Midha, P. S. 2001. A neural network approach for selection of powder metallurgy materials and process parameters. *Artificial Intelligence in Engineering* 14:39–44.

Dhakal, H. N., Zhang, Z. Y., and Richardson, M. A. W. 2007. Effect of water absorption on the mechanical properties of hemp fibre reinforced unsaturated polyester composites. *Composites Science and Technology* 67:1670–1683.

Goel, V., and Chen, J. 1996. Application of expert network for material selection in engineering design. *Computers in Industry* 30:87–101.

Herrera-Franco, P. J., and Valadez-Gonzalez, A. 2004. Mechanical properties of continuous natural fibre reinforced composites. *Composites Part A: Applied Science and Manufacturing* 35:339–345.

Idicula, M., Malhotra, S. K., Joseph, C., and Thomas, S. 2005. Dynamic mechanical analysis of randomly oriented intimately mixed short fibre reinforced polyester composites. *Composites Science and Technology* 65:107–1087.

Jayaraman, K., and Battacharyya, D. 2004. Mechanical performance of wood fibre-waste plastic composites. *Resources Conservation and Recycling* 41:307–319.

Joseph, S., Sreekala, M. S., Oommen, Z., Koshy, P., and Thomas, S. 2002. A comparison of the mechanical properties of phenol formaldehyde composites reinforced with banana fibres and glass fibres. *Composites Science and Technology* 62:1857–1868.

Lai, C. Y., Sapuan, S. M., Ahmad, M., Yahya, N., and Dahlan, K. Z. H. M. 2005. Mechanical and electrical properties of coconut coir fibre reinforced polypropylene composites. *Polymer-Plastics Technology and Engineering* 44:619–632.

Maleque, M. A., Belal, F. Y., and Sapuan, S. M. 2007. Mechanical properties study of pseudo-stem banana fibre composites. *Arabian Journal of Science and Engineering* 15:647–652.

Sapuan, S. M., and Abdalla, H. S. 1998. A prototype knowledge-based system for the material selection of polymeric-based composites for automotive components. *Composites Part A: Applied Science and Manufacturing* 29A:731–742.

Sapuan, S. M., Harimi, M., and Maleque, M. A. 2003. Mechanical properties of epoxy/coconut shell filler particle composites. *Arabian Journal of Science and Engineering* 28:171–181.

Sapuan, S. M., Jacob, M. S. D., Mustapha, F., and Ismail, N. 2002. A prototype knowledge-based system for materials selection of ceramic matrix composites of automotive engine components. *Materials and Design* 23:701–708.

Sapuan, S. M., Leenie, A., Harimi, M., and Beng, Y. K. 2006. Mechanical properties of woven banana fibre reinforced epoxy composites. *Materials and Design* 27:689–693.

Sastra, H. Y., Siregar, J. P., Sapuan, S. M., and Hamdan, M. M. 2006. Tensile properties of *Arenga pinnata* fibre reinforced epoxy composites. *Polymer-Plastics Technology and Engineering* 45:149–155.

Sastra, H. Y., Siregar, J. P., Sapuan, S. M., Leman, Z., and Hamdan, M. M. 2005. Flexural properties of *Arenga pinnata* fibre reinforced epoxy composites. *American Journal of Applied Sciences* (Special Issue):21–24.

Sergent, P. 1991. *Materials information for CAD/CAM*. Oxford: Butterworth-Heinemann.

Sha, W., and Edwards, K. L. 2007. The use of artificial neural networks in materials science based research. *Materials and Design* 28:1747–1752.

Smith, L. N., German, R. M., and Smith, M. L. 2002. A neural network approach for solution of the inverse problem for selection of powder metallurgy materials. *Journal of Materials Processing Technology* 120:419–425.

Tan, T. T. M. 1997. Thermoplastic composites based on jute fibre treated with cardanol formaldehyde. *Polymer and Polymer Composites* 5:273–279.

Tanvir, M. S., and Mujtaba, I. M. 2006. Neural network based correlations for estimating temperature elevation for seawater in MSF desalination process. *Desalination* 195:251–272.

Wambua, P., Ivens, J., and Verpoest, I. 2003. Natural fibres: can they replace glass in fibre reinforced plastics? *Composites Science and Technology* 63:1259–1264.

Yang, S. Y., Girivasan, V., Singh, N. R., Tansel, I. N., and Kropas-Hughes, C. V. 2003. Selection of optimal material and operating conditions in composite manufacturing: Part II. Complexity, representation of characteristics and decision making. *International Journal of Machine Tools & Manufacture, Design Research and Application* 43:175–184.

Yang, S. Y., Tansel, I. N., and Kropas-Hughes, C. V. 2003. Selection of optimal material and operating conditions in composite manufacturing: Part I. Computational tool. *International Journal of Machine Tools & Manufacture, Design Research and Application* 43:169–173.

Zhang, Z., and Friedrich, K. 2003. Artificial neural networks applied to polymer composites: a review. *Composites Science and Technology* 63:2029–2044.

Index

A

ABAQUS, 259, 271
Activation functions, 22, 41
  ANN in structural health
    monitoring, 42–44
Adam's equation, 211
Adaptivity, 40
Adhesively bonded composite single lap
    joint, 259, 261, 265
  chart, 272
  composite tubular single lap joint,
    271–286
  fibber volume effect, 262
  formation, 258
  free vibration analysis, 258–270
  micromechanics, 256–257
  model strain energies, 264
  modes, 286
  motion dynamic equations, 255
  optimal design, 251–290
Adhesively bonded composite
    structures
  beams multilayered, 253
  plates multilayered, 253
  shells multilayered, 253
  vibration analyses, 254
Adhesively bonded composite tubular
    lap joint, 251–290
  ANN model, 282
  bending modes, 288
  clamped-free, 288
  composite materials
    micromechanics, 256–257
  composite single lap joint free
    vibration analysis, 258–270
  composite tubular single lap joint,
    271–286
  fiber angle, 288
  fiber volume fraction, 288
  free vibration analysis and optimal
    design, 251–290
  inner tube radius, 288
  motion dynamic equations, 255
Adhesively bonded composite tubular
    single lap joint, 273

design parameters, 287
effect, 283
effect inner tube radius, 281
effect of outer tube thickness, 280
fiber angle effect, 275
fiber volume fraction, 276
inner tube radius, 284–285
inner tube thickness effect, 279
mesh detail of, 273
mode shapes, 274
natural frequencies, 274
overlap length effect, 277, 284–285
Adhesive materials
  mechanical properties, 273, 286
  physical properties, 273
Aging tests for cement-based
    composites, 165
Aircraft composite wing box structure
  hidden unit selection, 148–149
  impact estimation percentage error,
    156
  impact events applied to analyzed
    zones, 146
  impact events summary, 146
  impact positions, 145
  series of impacts, 144
  unit selection, 148–149
Anisotropic plates
  application, 123
  damage localization, 132
  novelty index plot, 131
ANN. *see* Artificial neural network
    (ANN)
ANNIE. *see* Artificial Neural Networks
    in Engineering Conference
    (ANNIE)
Artificial neural network (ANN), vii
  applications, 4–6, 160
  composite fatigue failure guidelines,
    212–214
  composite materials, 217
  composites, 4–6
  composition, 3
  data normalization, 213
  data sets for topology performance,
    173